RSGB

Building a Transceiver

by

Eamon Skelton, EI9GQ

and

Elaine Richards, G4LFM

Published by Radio Society of Great Britain of 3 Abbey Court, Priory Business Park, Bedford MK44 3WH, United Kingdom,

First Printed 2014

ISBN: 9781 9101 9301 3

Cover design: Kevin Williams
Typography and design: Mark Pressland
Production: Mark Allgar, M1MPA

Printed in Great Britain by Latimer Trend & Company Ltd of Plymouth

Contents

Introduction

CONVENTIONAL WISDOM says that a home-made rig is easy to repair: if you make it yourself, you should be able to fix it yourself. This relatively complicated HF transceiver project is from the Homebrew series in RadCom written by Eamon Skelton, EI9GQ and has been broken down in to modules that can be built and tested individually. Each module will be described in cookbook fashion so that the constructor understands how it works before starting the build. The constructor may choose to build a complete transceiver based closely on this design, add in some of your own modules or you may prefer to mix and match by using parts of this design and integrating them into a completely different project. The choice is yours.

In addition to providing all the details you need to build a top quality amateur radio transceiver, Eamon talks you through his design thinking. The discussion covers the design options he considered along with his reasons for choosing the selected circuit. The discussion provides a valuable insight into the design process itself and vital information for those wanting to deviate from the original design. Reading through Eamon's descriptions will help you better understand the inner workings of a modern amateur transceiver and provide vital information should you need to do any fault finding, either to complete the project or effect repairs at a later date.

Eamon's modular approach to the transceiver design also makes it very easy to extract sections for use elsewhere. A good example is to be found in the VFO design that could easily be adapted for use as a signal source for other projects or test equipment. The broadband RF amplifiers are similarly useful. Eamon's modular approach also means that each module can be tested in isolation, therefore giving a much greater chance of success when the full transceiver is interconnected and powered-up. This approach also makes fault finding very much simpler as you can quickly isolate any fault to a module before needing to delve into component level fault-finding.

The project starts with the transmit section working through the generation of SSB signals to the mixers and band pass filters needed. This is followed by the amplification stages creating a signal suitable for transmission. The receiver stage starts with the band pass filter before the signals reach the mixer stage. This is followed through until the audio amplifier stage when we are ready to look at the band switching needed to connect the projects together. At the end of the project are a couple of power amplifiers, one for 200W and one for 400W, to make your finished project capable of transmitting at a level suitable for your station.

The testing and measurement of HF radio equipment often calls for the

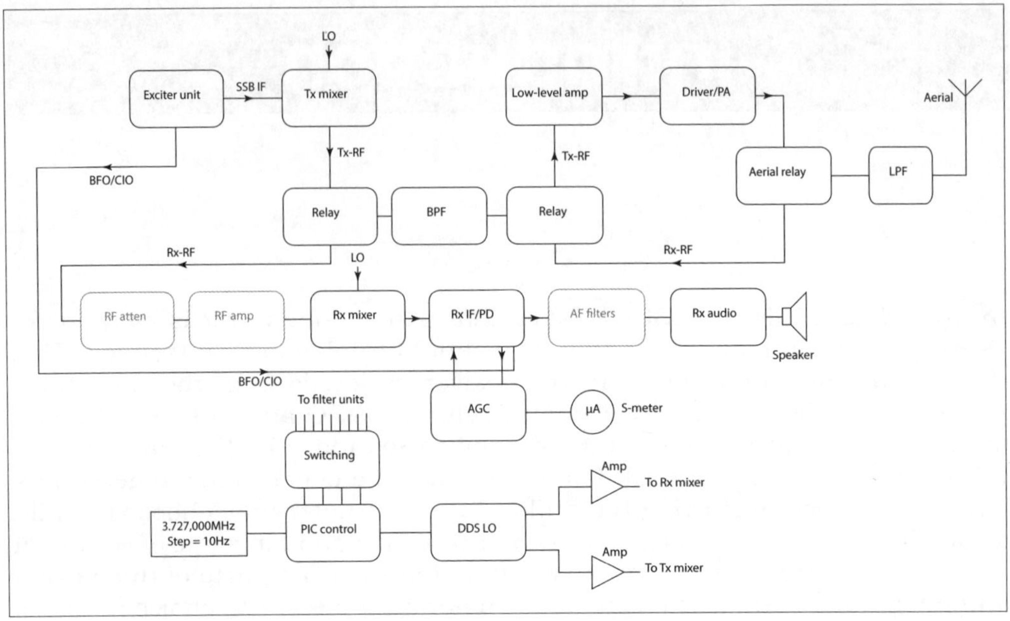

use of specialised test equipment that is outside the budget of most hobbyists. Eamon offers a different approach in this project and describes the design and use of a number of very simple but accurate measurement tools and techniques. These tools would make a valuable addition to any shack. Eamon's description of the test tools will also help you better understand the thinking behind each measurement technique and aid your interpretation of the results.

In addition to covering the detailed electronic design process, Eamon also provides extensive details of the practical constructional techniques necessary to build the transceiver. Eamon is famous for his dead-bug construction style and this is fully covered along with a number of other useful techniques. Photographs are also provided to help the reader visualise the final layout of each section. All Eamon's construction techniques are very simple and use the tools and facilities that are likely to be found in most amateur shacks.

In summary, Eamon's cookbook style creation of a modern HF transceiver not only provides a high quality project but also educates the reader in the design process. You may never build the complete transceiver described in this book but the construction techniques and testing has been designed with the resources of an amateur radio shack in mind, so will be useful whatever you decide to build.

1

A Simple SSB Exciter Oscillator & Balanced Modulator

THIS PROJECT is a conventional filter method SSB generator. **Photo 1** shows the prototype SSB exciter and **Figure 1** shows a simplified block diagram. The carrier is generated by a pair of crystal oscillators. A couple of buffer amplifiers increase the carrier signal to a level that is suitable for driving a simple diode double-balanced modulator (mixer). The AF signal from the microphone is amplified by a simple opamp based amplifier/LPF and then applied to the other input port of the balanced modulator (BM). The balanced modulator is a modulator that suppresses both the carrier and the AF input from the microphone amplifier, leaving only the sum and difference frequencies at its output. These are the upper sideband (USB) and lower sideband (LSB) signals. This double-sideband signal is passed through a very selective IF filter that removes the unwanted sideband. The IF

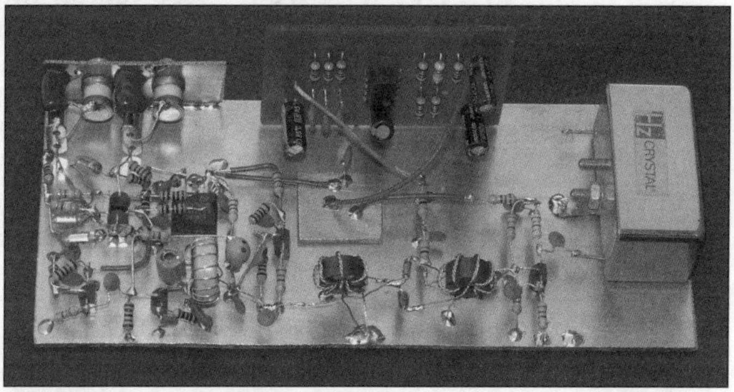

PHOTO 1: The SSB exciter. The oscillator section is on the left, balanced modulator bottom-centre and the crystal filter on the right. The PCB at the top is a mic amplifier.

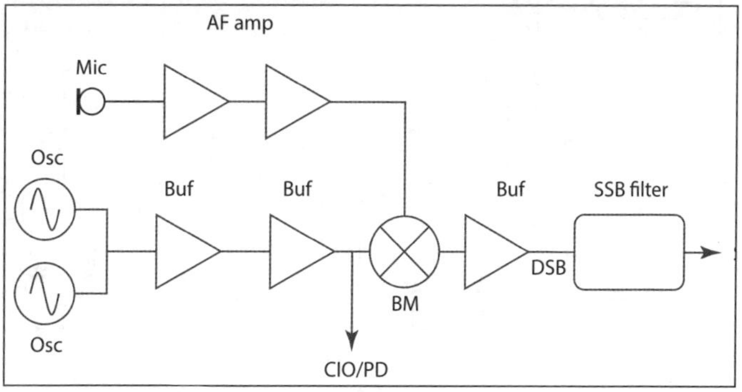

FIGURE 1: Block diagram of the SSB exciter.

filter further attenuates the carrier that was suppressed in the BM.

Oscillators and Buffer Amplifiers

To keep USB/LSB switching as simple as possible, I used two separate crystal oscillators to generate the LSB and USB carriers. It would have been possible to use a single oscillator and a simple electronic switch or a relay to select the correct crystal for the required sideband mode. The extra cost and complexity of the dual oscillator configuration is easily justified by the greatly simplified USB/LSB switching and fact that no extra transistors, PIN diodes, MOS switches or other temperature sensitive components will be placed in circuit with the crystal resonators. This scheme is easily extended to accommodate a third oscillator with a suitable offset for CW reception if this is required.

The oscillators were designed with good frequency stability as a primary consideration. The oscillator and first buffer amplifier are powered from a 6V stabilised DC supply. The capacitors in the frequency-determining parts of the oscillators are high stability polystyrene types. Silvered mica or NPO type ceramic capacitors should also perform well in this circuit. The oscillator output is buffered by an emitter follower (common-collector) amplifier. This has no voltage gain, but it provides good isolation between the oscillators and the final carrier amplifier. The final amplifier is a simple common-emitter amplifier. This is followed by a simple pi low-pass filter network designed for a collector load impedance of

$$X_{C1} = \frac{R1}{Q_L}$$

$$X_{C2} = R_L \sqrt{\frac{\frac{R_1}{R_L}}{(Q_L^2 + 1) - \frac{R_1}{R_L}}}$$

$$X_{L1} = \frac{Q_L R_1 + \frac{R_1 R_L}{X_{C2}}}{Q_L^2 + 1}$$

FIGURE 2: Standard handbook formula for determining the oscillator output filter network.

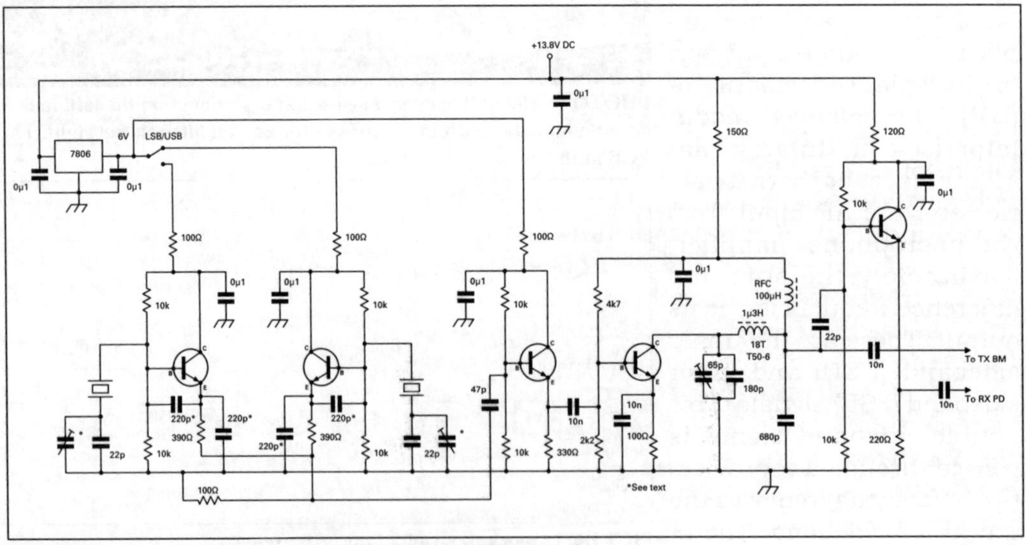

FIGURE 3: Circuit diagram of the SSB exciter oscillator section.

450Ω, output impedance of 50Ω and a loaded Q of 7. The standard handbook formula **[1]** for designing this network is shown in **Figure 2.**

Using:
R1 = 450Ω,
R2 = 50Ω,
QL = 7,
f = 10.7MHz
gives us reactance values of XC1 = 64.285 (231pF), XC2 = 23.426 (635pF) and XL2 = 82.209 (1.223μH).

These values were tweaked slightly to fit the nearest available component values. The oscillator section schematic is shown in **Figure 3.**

Oscillator Construction

The exciter was built on a strip of copper PCB laminate. All five transistors in the oscillator section are BC548 or similar general purpose NPN types. I used a 10.7MHz IF for testing the exciter. This is not necessarily my final choice of IF for the HF transceiver. The centre frequency of my IF filter is 10.6985MHz. The USB carrier oscillator is 1.5kHz lower at 10.697MHz and the LSB carrier oscillator is at exactly 10.700MHz. Each crystal is tuned to resonance by a series capacitance. The exact value of this capacitance will depend on the characteristics of the crystals used. A value of around 30pF is typical. I used a 22pF silvered mica capacitor in parallel with a high quality 12pF air dielectric trimmer capacitor. You may need different values to match your crystals. The four 220pF feedback capacitors in the emitter circuits of the oscillators are polystyrene types. A 7806 voltage regulator provides the 6V DC supply for the oscillators and the first buffer amplifier. As the current drawn by the oscillator and first buffer is only a few mA, it is not necessary to provide a heat sink for the regulator IC. The 100nF decoupling capacitors are ordinary disc ceramic types.

High quality components and good VHF construction methods will help you to achieve the best possible oscillator stability. The choice of components for the buffer amplifiers is not as critical. Ordinary ceramic capacitors can be used in these stages. The 1.3μH inductor in the output network is 18 turns of 0.375mm enamelled wire (see later) on a T50-6 (yellow) powdered iron toroid. A 65pF trimmer capacitor (Philips yellow or similar) is used as part of C1 in the pi network. The 100μH RFC is a miniature moulded type. The value of this component is not critical; any RFC with a value of 22μH to 220μH would make a suitable replacement.

Note that the oscillator section has a high level output to the balanced modulator and a separate carrier output via an emitter-follower buffer amplifier. This output can be used to provide a beat frequency oscillator / carrier injection oscillator (BFO/CIO) signal for the receive section of a transceiver. This

PHOTO 2: Oscillator section of the exciter.

stage is optional and it can be omitted if you are just building a transmitter. **Photo 2** shows the oscillator section of the exciter. The circuit might look like a bit of a rat's nest, but it is electrically and mechanically very stable. At the left side of the photo you can see how the crystals and their trimmer capacitors are mounted on a small strip of double sided PCB that is soldered to the main PCB ground plane. Both crystals were pre-tinned with solder at one top corner and then soldered to the vertical PCB. If you decide to use the same approach, don't apply the iron to the crystal mounting can for more than a few seconds or you may cause damage, including melting the can's soldered hermetic seal.

Oscillator Testing

The oscillator output was terminated with a 50Ω resistor and an oscilloscope was used to measure the output voltage. The 65pF trimmer in the output network was adjusted for maximum voltage across the output load. The two oscillators were adjusted to the correct frequencies of 10.697MHz and 10.7MHz using a frequency counter. Once final testing is complete, the oscillator frequency can be set to the exact value required after a suitable warm-up period. There was a slight difference in output level between the two oscillators, but the measured output was between 2.2Vp (4.4Vp-p) and 2.5Vp (5Vp-p) for both oscillators. This is around 50-60mW (or +17dBm), which is an ideal level for driving the LO port of a high level diode double-balanced mixer (DBM). Frequency stability seems to be excellent. The measured drift is within 1Hz per hour after a few minutes warm-up. The greatest frequency error measured over a period of a few days was less than 10Hz. This shows medium to long term stability of 1ppm, which is about as much as can be achieved by a simple crystal oscillator. I will have to wait for a few cold winter mornings for the final stability tests. The output signal from the oscillator is a nice, clean and symmetrical sine wave. Measured second harmonic output is -45dBc. No other spurious outputs were detectable.

Balanced Modulator

After considering the various options available, I decided to use a simple diode DBM as the balanced modulator. This simple device offers much better performance than any of the commonly available balanced mixer/modulator ICs like the NE602/612, SA602/612 etc. High level mixer/modulator ICs are available, but they are not that easy to find and they are usually quite expensive. Such devices offer performance that is almost as good as, but certainly not better than the humble diode mixer. The disadvantages of the diode mixer are the requirement for a high level of local oscillator injection (carrier oscillator in this case) and a small conversion loss instead of the conversion gain of a typical active mixer. The high LO injection level requirement is easily met by our new carrier oscillator unit. A simple transistor amplifier will compensate for the mixer conversion loss. The price/performance ratio of the diode mixer/modulator is absolutely unbeatable. Four diodes will cost between 4p and 40p depending on diode type. Junkbox toroids usually come at zero cost. The ferrite ‘cups’ from Toko and similar types of 10mm RF/IF transformers make excellent toroid cores. Even if you buy a pair of FT37-43 toroids, you can make a high performance BM for well under £1.

For many years, it was normal practice to tolerate a fairly high level of distortion of the signal from the balanced modulator. All out-of-passband spurious signals caused by harmonics of the AF signal from the microphone amplifier and odd-order IMD products (which are the most common cause of ‘splatter’) are very effectively removed by the SSB filter. Many SSB exciters use AF compression and/or clipping of the modulation signal, RF/IF compression and/or clipping of the IF signal from the balanced modulator and non-linear variable gain IF amplifiers or attenuators as a method of applying ALC (automatic level control). The distortion generated by these various devices will affect the quality of signals within the SSB filter passband, but should not cause a significant increase in spurious signals outside the filter passband. In some cases, a small amount of clipping and/or compression might be considered desirable because it increases the average output power of the transmitter.

Modern trends call for improved linearity of the transmit AF and IF systems. Improved SSB quality will result in cleaner signals and less splatter on the bands. The proliferation of narrow bandwidth modes like PSK31 places greater emphasis on

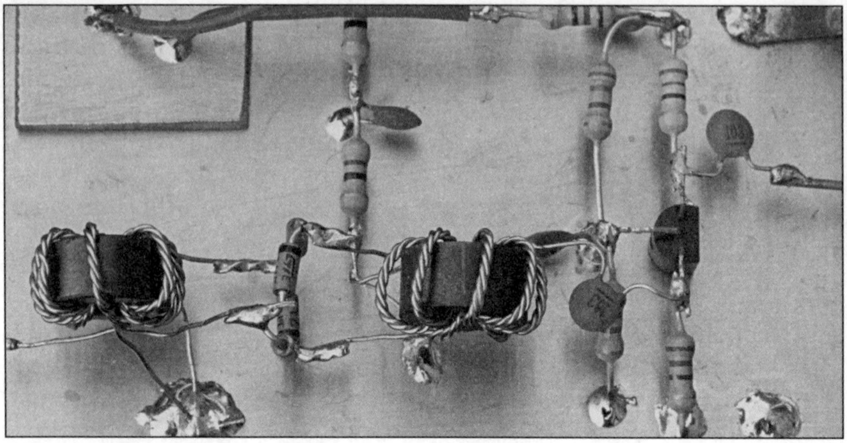

PHOTO 3: The balanced modulator.

transmitter linearity. As it is possible to fit more than a dozen PSK31 signals within the passband of a typical SSB filter, it will be obvious that poor in-passband linearity that does not produce splatter outside the SSB filter passband could still cause severe interference to other PSK31 stations on adjacent frequencies.

Balanced Modulator Construction

Figure 4 shows the schematic of the balanced modulator and amplifier. The modulator is a standard diode mixer based on a ring of 1N5711 Schottky diodes. T1 and T2 are each 7 turns of trifilar wound 0.375mm enamelled copper wire (Maplin YN86T or similar). Any enamelled wire with a diameter of 0.3-0.5mm should be equally suitable. The toroid cores are FT37-43 or similar medium to high permeability ferrite toroids. I have built several similar mixers and found that all of them have very high LO to RF port isolation of around 50dB, so I have not included any carrier balance adjustment in my prototype circuit. More about this later...

The BM is followed by a simple common-base transistor amplifier. This amplifier acts as a buffer between the mixer output port and the highly reactive input impedance of the crystal SSB filter. The amplifier also provides the required impedance transformation from the 50Ω mixer output port up to several hundred ohms at the input of the SSB filter.

When a DBM is used as a balanced modulator, the ports are configured so that the IF port becomes the AF input and the RF port becomes the IF output port. The mixer IF port provides the only input that can operate at frequencies from VHF all the way down to DC. The simple RC network at the mixer IF port provides a proper termination for RF signals. I used 47Ω resistors because this was the nearest available standard value to 50Ω. You can, of course, use 51Ω resistors if they are available, or simply put two 100Ω resistors in parallel. The 1N5711 diodes were matched for forward voltage drop using a digital multimeter. Other Schottky

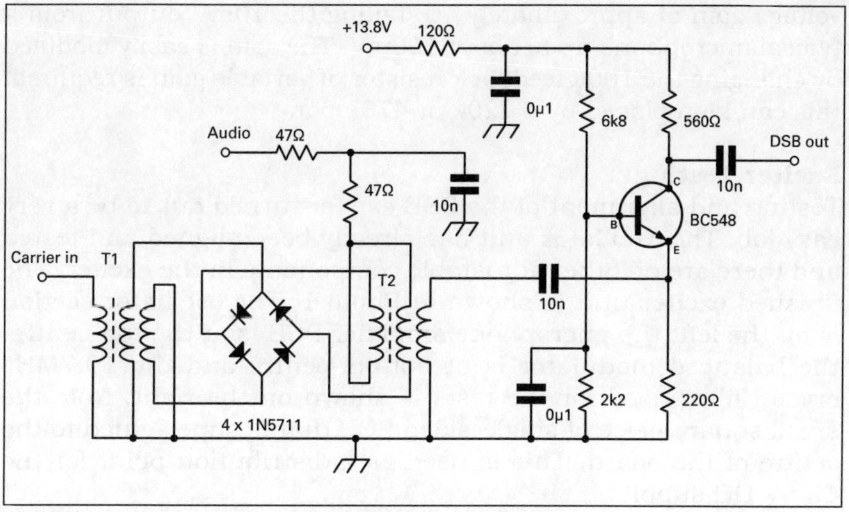

FIGURE 4: Circuit of the balanced modulator and amplifier.

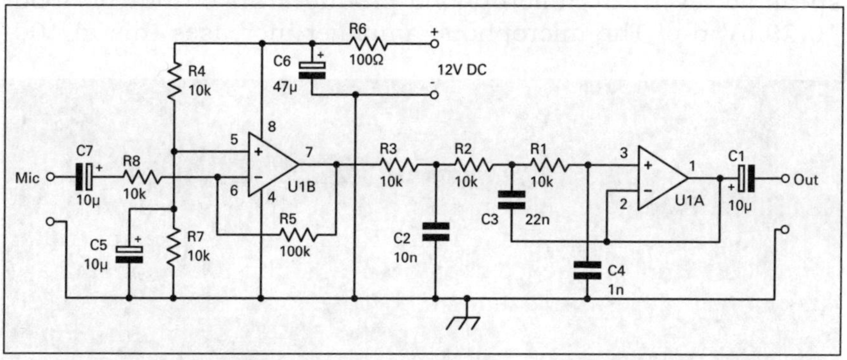

FIGURE 5: Microphone amplifier.

diodes like the BAT43 (Maplin VR19V) should also work well in this circuit. I haven't tried using ordinary switching diodes like the 1N914 or 1N4148 in this circuit, but I would expect them to work well. **Photo 3** shows the balanced modulator. Note how T1 and T2 are made. The 7 trifilar turns are evenly spaced around the toroid core. The three strands of enamelled copper wire are twisted tightly together so that the three windings are physically and electrically identical. The four diodes are formed into a diamond shape with the leads twisted together and then soldered. This uncharacteristically neat, tidy and symmetrical construction gives my prototype remarkably good carrier suppression. I used a BC548 transistor for the common-base amplifier.

Microphone Amplifier

The microphone amplifier of **Figure 5** was first described in the August 2006 Homebrew. It can be built dead-bug style like the other stages, or if you prefer to make a PCB, the artwork can be found at http://tinyurl.com/2wb7aut. The microphone amplifier has a voltage gain of approximately 10, taking the 10mV output from a typical microphone up to about 100mV. The gain is easily modified by changing the 100k feedback resistor; if variable gain is required, this can be replaced by a 220k or 470k pot.

Exciter Testing

Testing and alignment of the SSB exciter turned out to be a very easy job. The oscillator unit has already been aligned and tested and there are no other adjustable components in the exciter. The finished exciter unit is shown in Photo 1. The oscillator section is on the left, the microphone amplifier PCB is at the top-centre, the balanced modulator is at bottom-centre and the 10.7MHz crystal filter used for the tests is shown on the right. Note the small square piece of single-sided PCB that is super-glued to the centre of the board. This is used as a distribution point for the 13.8V DC supply.

I used my desk microphone for the initial tests. At a normal speaking level, this microphone produces an output of about 10-20mV p-p. The microphone amplifier increases this to 100-

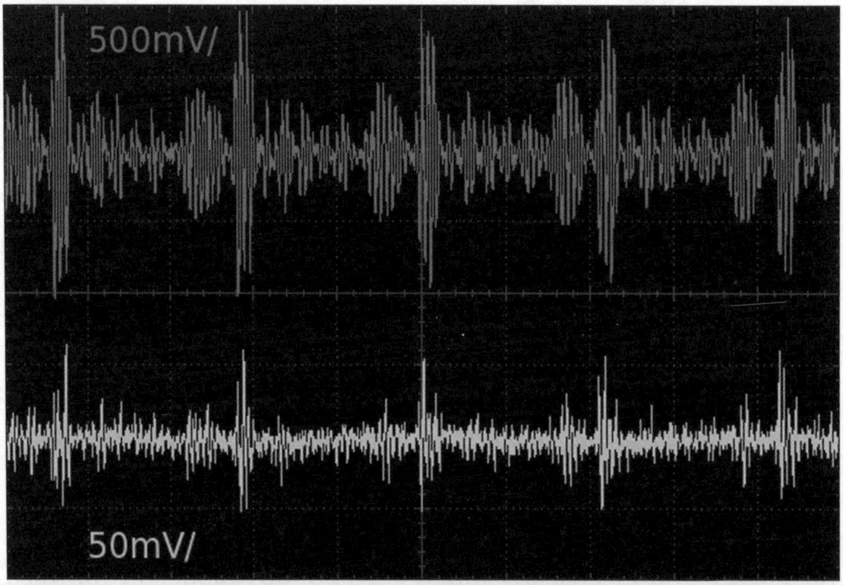

FIGURE 6: SSB exciter input (bottom trace, 50mV/div) and output (top trace, 500mV/div).

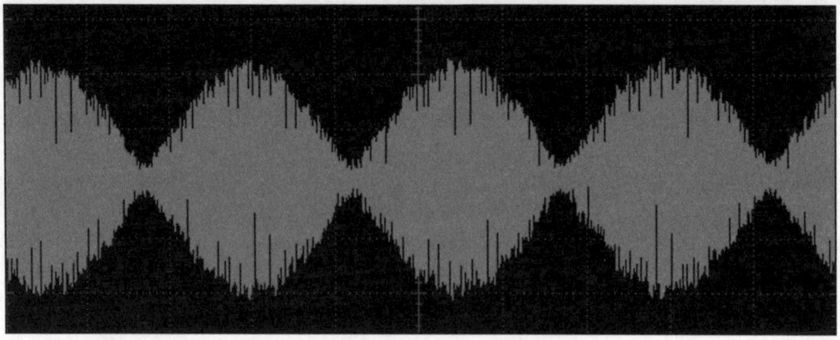

FIGURE 7: Two tone test signal output from the exciter.

200mV p-p (or 70.7mV RMS). This is -10dBm, or 100 microwatts into 50Ω. Applying a few well-proven rules of thumb tells us that the mixer/modulator output third order intercept point will be approximately the same as the oscillator injection level of +17dBm. The input IP3 will be about 6dB higher at +23dBm, 33dB above the modulation audio level. This should result in a very clean SSB signal with third-order IMD some 66dB below the desired output for in-band IMD. Distortion products outside the SSB filter passband should be so low as to be unmeasurable.

The measured voltage gain of the common-base base amplifier at 10.7MHz is 18-20x. At first glance, this would seem to indicate a gain of 26dB. The real power gain is much less because the amplifier output load impedance is higher than the input impedance. However, the overall circuit has considerable gain. 10mV at the microphone input produces several hundred mV of SSB at the filter output. **Figure 6** shows oscilloscope measurements of the input (lower trace at 50mV/div) and output (upper trace at 500mV/div) of the common-base amplifier. **Figure 7** shows the output signal during a two-tone test. An audio sample of the first test can be found at http://homepage.eircom.net/~ei9gq/exc_test.mp3. The audio quality seems quite good, although I find it a little too bassy. This is probably due to the characteristics of the microphone. As expected, carrier suppression at the output of the common-base amp is better than 50dB and more than 60dB below peak output when measured at the output terminal of the SSB filter.

A Few Tips

It is possible to add carrier balance adjustment to a diode DBM. Home made mixers have an advantage over pre-packaged mixers in this regard. It is a lot easier to add the balancing components when the mixer is not hidden in a tin can. **Figure 8** shows how a carrier balance pot can be added to the BM. Ideally the resistance of the potentiometer should be quite low relative to the mixer termination impedance of 50Ω. As low value pots are not easy to find, I used a potentiometer made from a combination of one variable pot and two fixed resistors. This circuit also includes a pair of low value capacitors that can be used to tune

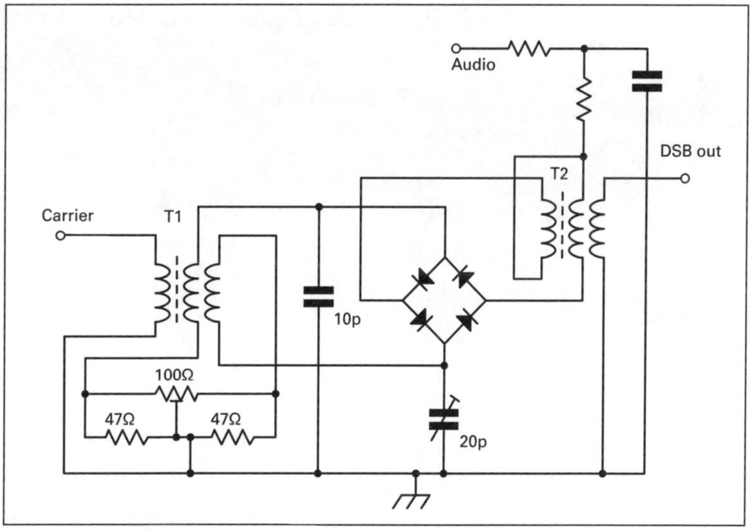

FIGURE 8: Adding a balance adjustment to the balanced mixer.

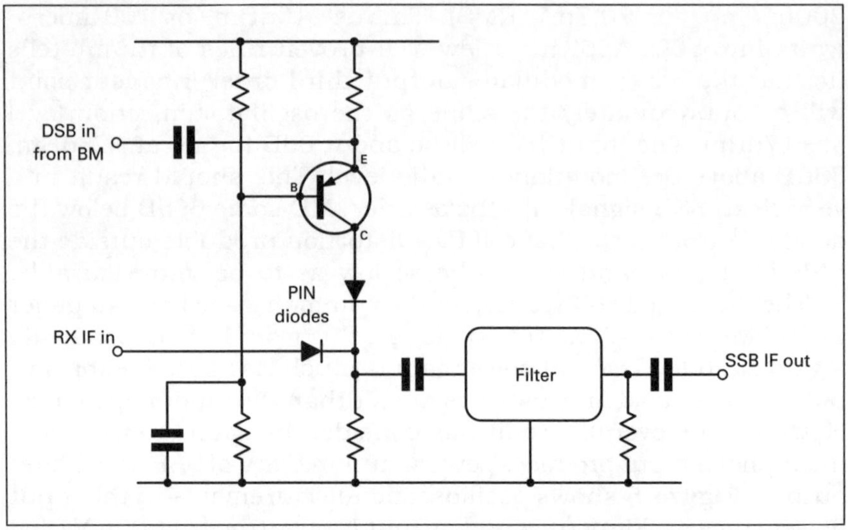

FIGURE 9: A simple but effective way to switch the SSB filter for transmit and receive

out any imbalance due to stray reactances in the circuit.

Figure 9 shows a simple but effective switching method that allows the same SSB filter to be used for transmitting and reception. This is achieved by using a PNP transistor in the common-base amplifier so that the circuit is effectively turned upside-down. The filter termination resistor becomes the collector load resistance. A PIN diode switch disconnects the amplifier from the filter when the switch is biased for receive operation. Be careful when you are experimenting with DC bias currents for diode switching. DC current applied to the I/O pins of a crystal filter could easily burn out the very thin wire used for the internal transformers.

[1] *ARRL Electronics Data Book*, ARRL 1976

2

Transmit mixers and band pass filters

THIS PROJECT uses several NPN transistors. The stages that we have built so far have been based on the well known and widely available BC547, 548 and 549 series of transistors. Transistors used in broadband amplifiers should have an fT (unity gain transition frequency) that is many times the upper frequency limit of the amplifier. To ensure that the broadband amplifiers described here will have good performance from 1.8MHz to 52MHz, I will use VHF/UHF transistors in these stages. This will leave me with the option of covering the 6m band if I decide to add an extra set of RF filters at some time in the future. Many of our previous projects have used some of the old favourites like the BFY90, BFR91, 2N5179, 2N2369A etc. These devices are still available, but they are now quite old and prices tend to be high. After trawling through the data sheets and price lists, I have chosen the MPSH10 as the new standard Homebrew VHF transistor. This transistor is currently in full production and available in both leaded and surface-mount packages. It is RoHS compliant and can be bought in bulk for about 7p per device. The MPSH10 is specified as a low noise UHF/VHF amplifier **[2]**.

The SSB exciter described previously is capable of producing a high quality SSB IF signal. This intermediate frequency signal is passed through a crystal filter that does a very good job of suppressing the opposite sideband and any other out-of-passband signals. At this point in the circuit, the signal quality is very high. Everything we do in the following stages will degrade the signal quality to some extent. Our task is one of damage limitation. We must convert this milliwatt level IF signal to a different frequency within one or more of the amateur bands, filter out and suppress any spurious signals that are generated in the frequency conversion process and then amplify the low level RF signal up to a level of several tens or possibly even several hundred watts.

Any harmonics generated in the transmitter amplifier stages are easily removed by a simple low pass filter (LPF) between the PA stage and the aerial. Other distortion products are not so easy to remove. Odd-order intermodulation distortion (IMD) products are particularly problematic because they tend to produce spurious signals at frequencies that are close to the transmit frequency. To keep distortion to an acceptably low level, all of the transmitter stages beyond the IF filter must have excellent linearity.

As the intermediate frequency (IF) signal is at a very low power level, it would be possible to use a low level mixer for the IF to RF conversion. Many published designs use active IC mixers for this application. This approach has the advantage of simplicity but does not offer the best performance. To meet my requirement for excellent linearity, I have decided to use a high level passive diode mixer as the transmit mixer. A previous rig of mine used a conventional diode double balanced mixer (DBM) as the transmit mixer. This has served me very well for many years. The DBM offers excellent local oscillator (LO) to RF and LO to IF isolation of 40-60dB. This makes it a lot easier to keep the very strong LO signal out of the transmit RF amplifier stages. The IF to RF (and vice versa) port isolation of a typical DBM is not quite so good: typical values range from 15-25dB. This could lead to problems where the IF is close to the intended output frequency. Poor suppression of a 10.7MHz IF would be likely to cause problems in a 10MHz (30m) transmitter. A 40m transmitter with a 7.8MHz IF is another tricky combination.

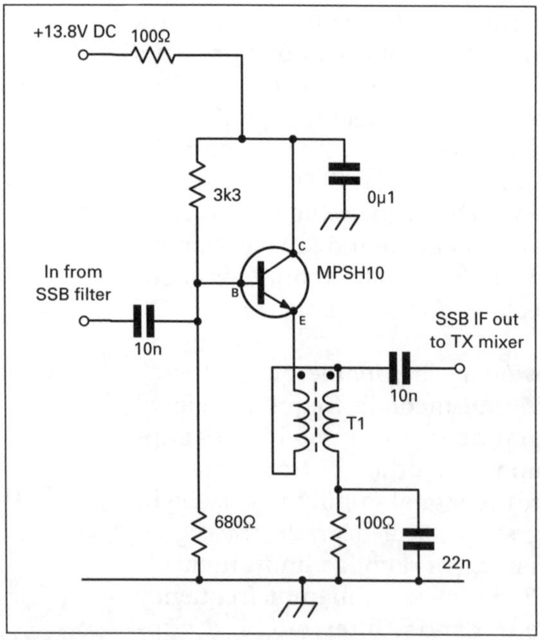

FIGURE 10: Buffer amplifier between the SSB filter (exciter unit) and diode mixer IF port.

Like most home constructors, I have a long list of things I would like to build. I will eventually get around to making some of them; others will probably stay on the to-do list forever. One interesting device that I have never built before is the triple balanced mixer (TBM) or, if you prefer, the double-double balanced-mixer. A typical TBM consists of a pair of diode rings. This configuration offers good isolation between all three ports. The circuit is a bit more complex than the standard DBM. Eight well-matched diodes and as many as five ferrite cores may be required. As the local oscillator power is switching twice as many diodes, the TBM requires more LO drive than the more common DBM. Another disadvantage of the TBM is that the IF port is not DC coupled. This means that it is not suitable for use as a phase detector or AF product detector. The advantages of the TBM are a high third order intercept point and good IF/RF port isolation. This should result in greater dynamic range and fewer spurious signals at the mixer output.

We will need to find a way of matching the IF filter termination impedance (usually around 500Ω) to the 50Ω IF port of the diode mixer. A simple emitter-follower buffer amplifier is one of the easiest ways of making a high to low impedance transformation. **Figure 10** shows the buffer amplifier that I used between the output of the SSB filter (exciter unit) and the IF port of the diode mixer. T1 is a 4:1 impedance ratio (2:1 voltage ratio) transformer made from 10 turns of enamelled wire, trifilar wound on a FT37-43 ferrite toroid. As the voltage gain of an emitter-follower is approximately 1, the transformer reduces the overall gain to 0.5. However, this amplifier has a small power gain, about 4dB, because the input impedance of 500Ω is higher than the output impedance. For instance, if the input voltage is 1V (RMS), the input power is $V^2/R=1/500=0.002W$ (or 2mW). The output power is $0.5^2/R=5mW$. The circuit was built dead-bug style using a MPSH10 transistor.

Triple Balanced Mixer

I have several home-and commercially-made DBMs gathering dust on the shelf. I will do a few tests on a home-made triple balanced mixer before I decide whether I should use the TBM or stick with the easier and well tried option of using a DBM. **Figure 11** shows my prototype mixer. I used two rings of well matched

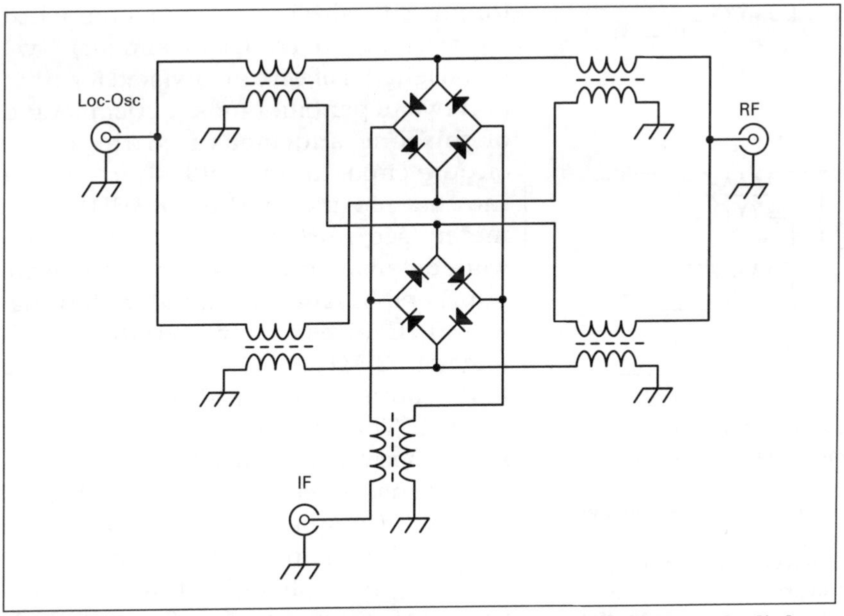

FIGURE 11: Prototype mixer using two rings of well-matched 1N5711 Schottky diodes.

1N5711 Schottky diodes. I considered a few different options for the mixer I/O balun transformers. **Figure 12** shows a few of the many configurations available. A is a conventional transformer with an optional centre tap on the secondary. B is a simple 1:1 transmission line transformer, often described as a 'sortabalun' in American publications. C is another 1:1 transmission line transformer with an extra winding to improve voltage balance. D is a 1:4 balun made from a pair of transmission line transformers. This configuration is often described as a 'Guanella balun'. Two separate ferrite cores are usually required, although it is possible to wind the two separate transmission lines on a single two-hole 'binocular' type of balun core. Previous experience has shown that transmission line transformers can operate over a wider bandwidth than conventional flux-coupled transformers. As I didn't have suitable binocular type cores for D, the two remaining options were B and C. After a few failed attempts at using balun transformers based on C, I eventually decided to use transformers based on Figure 12B. This is a miniature version of the 'choke balun' that is often used in aerial feedlines. A typical choke balun consists of several metres of coaxial cable wound on a plastic drainpipe. My miniature version is twelve turns of bifilar wound wire on a T37-43 ferrite toroid. The wire is 0.375mm enamelled copper (Maplin YN86T or similar). Two equal lengths of wire are twisted together at 2 twists per cm. I took a 20cm length of this line and measured the open-circuit capacitance and short-circuit inductance using the VK3BHR L/C meter (see *RadCom* May 2008 if you want details). The measured values were 90nH and 31.2pF. The standard formula $Z0=\sqrt{(L/C)}$ gives a characteristic impedance of 53.7Ω. This is reasonably close to the industry standard I/O impedance of 50Ω. The first (quarter wavelength) resonance was measured at 190MHz using a signal generator, RLB and 50Ω load resistor. Comparing the physical line length (20cm) to a free-space quarter wavelength (75/f) shows that the velocity factor is just over 0.5. This is a little lower than I expected. Testing several

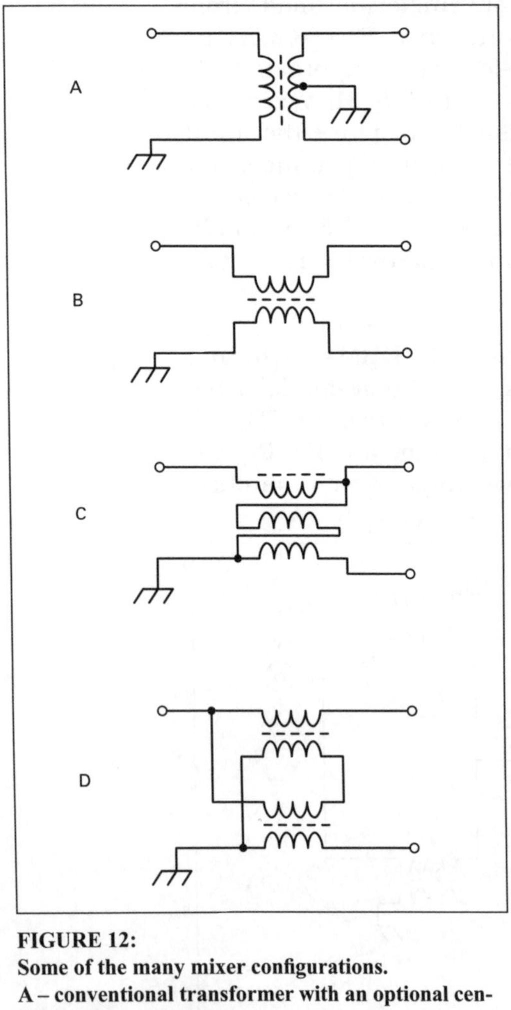

FIGURE 12:
Some of the many mixer configurations.
A – conventional transformer with an optional centre tap on the secondary.
B – simple 1:1 transmission line transformer or 'sortabalun'.
C – another 1:1 transmission line transformer with an extra winding to improve voltage balance.
D – 1:4 balun made from a pair of transmission line transformers, often known as a 'Guanella balun'.

PHOTO 4: The assembled triple balanced mixer with BNC terminations.

different line lengths for resonance showed some small variation in VF that suggests that the relative permittivity of the wire insulation is frequency dependant. However, in every case, the VF was always between 0.50 and 0.57.

Each of the five transformers in the TBM is 12 turns on a small high permeability ferrite toroid (FT37-43 or similar). As stated, the characteristic impedance of the line through the transformer is close to 50Ω. Line loss for normal mode signals is extremely low at MF/HF, rising to just 1dB at 200MHz. Because it is wound on a ferrite core, the line presents a very high impedance to common mode current. 12 turns on the FT37-43 has an inductance of 50.4μH: an inductive reactance of 570Ω at 1.8MHz, rising to more than 9kΩ at 29.7MHz. This large common mode to normal mode impedance ratio guarantees good current balance at the balanced end of the line.

Photo 4 shows the assembled mixer. Take care with the polarity of the diodes and the transformer windings. I took reasonable care with mine and I still managed to get one of the transformers wrong at the first attempt.

Performance

Testing the LO-RF and LO-IF port isolation with a signal generator and sensitive power meter shows that the performance of this mixer compares very well with a conventional DBM. Isolation is better than 50dB from 6.5MHz to well over 100MHz. Isolation falls to 39dB at 1.8MHz due to the reduced reactance of the transformer windings at LF. The measured RF-IF isolation is also in the 40-60dB range from Top Band to VHF. IF-RF isolation at the most common IF frequencies in the 8-12MHz range is 54dB. This is a substantial improvement over a conventional DBM. Measured IF to RF conversion loss is around 6.5dB. The tests were performed with 22.5mW (+13.5dBm) of local oscillator injection and 0.1mW (-10dBm) of IF injection.

Post Mixer Amplifier

The signal from the mixer RF port is amplified by a 20dB broadband amplifier, which is followed by a 4dB attenuator pad to give an overall gain of 16dB. The amplifier is based on a parallel pair of MPSH10 transistors. As this amplifier is placed before the RF band pass filters, it is critically important that it has very good

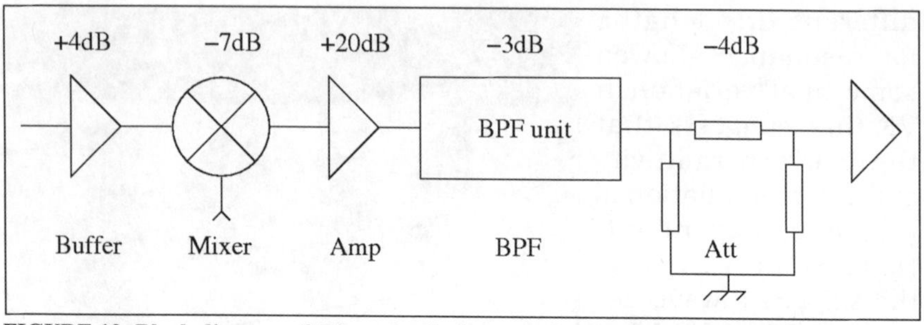

FIGURE 13: Block diagram of this project's elements showing distribution of gain and loss.

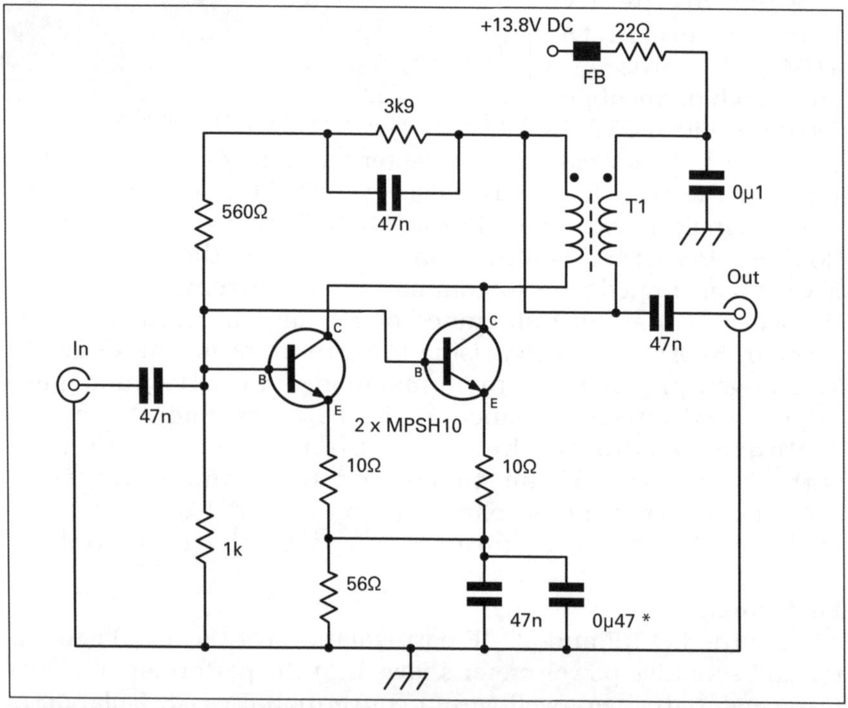

FIGURE 14: The post mixer amplifiers is extremely linear and gives about 16dB of gain.

linearity. **Figure 13** shows how the amplifier fits into the chain we're building. The amplifier schematic is shown in **Figure 14**. The amplifier was built on a strip of copper PCB laminate. T1 is 10 turns bifilar wound on a FT37-43 toroid. The base and collector leads of the two transistors are soldered together. The 5Ω degenerative feedback resistor is distributed equally between the two transistors by putting a 10Ω resistor in the emitter circuit of each transistor. As well as providing RF negative feedback, the resistors encourage equal current sharing between the two transistors. **Photo 5** shows the double-decker arrangement of the transistors.

The strongest signals at the amplifier input are the wanted RF signal and the IF image. The local oscillator signal is much reduced because of the LO-RF isolation of the mixer, but it is still relatively strong at about 15-20dB below the RF signal.

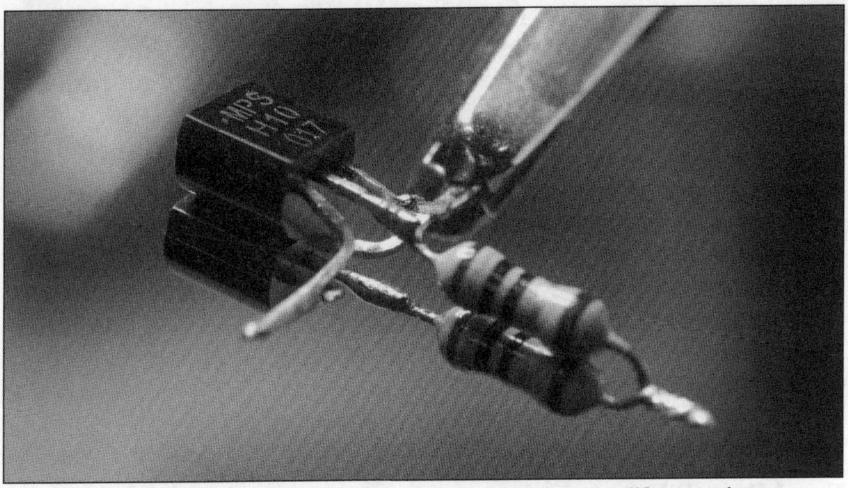

PHOTO 5: The double-decker arrangement of the post mixer amplifier transistors.

The first prototype had a measured gain of just over 20dB, falling to 19dB at 0.9MHz and 73MHz. The 3dB points were at 0.5MHz and 102MHz. This meets or exceeds our design requirements. Output return loss (RL) was measured at better than 20dB (SWR=1.2:1) from 1.5MHz to 87MHz. Input return loss was a slightly disappointing -14dB (SWR=1.5:1) from 2.1MHz to 120MHz. I have used several 47nF capacitors in the amplifier instead of my usual choice of 100nF. 47nF is just about ideal for the input/output coupling capacitors. The 47nF capacitor used to bypass the 56Ω resistor in the emitter circuit has a reactance of 1.88Ω at 1.8MHz. This is relatively high when compared to the value of the 5Ω degenerative feedback resistance. Putting a 470nF capacitor in parallel with this capacitor improved the input RL at low frequencies. The modified amplifier shows input RL of 14dB from 1.2MHz to 177MHz and 16dB (SWR=1.4:1) from 3.5MHz to 70MHz). The prototype RF amplifier is shown in **Photo 6**. Note that the extra 470nF capacitor is not

PHOTO 6: The prototype broadband RF amplifier.

shown in this photo. So far I have built three of these amplifiers: one as the post mixer amplifier and one as part of my test rig for evaluating the performance of the circuits made so far. The third has been boxed up as a general purpose 20dB amplifier.

As an experiment, I replaced the transistors with a pair of BC548s. This reduced the amplifier bandwidth to less than 30MHz. I then substituted a single BFG135. This is a surface mount UHF 1W transistor with an fT of 7GHz. I was surprised to find that this transistor did not give significantly wider bandwidth than the MPSH10 pair. Obviously, transistor fT is not the only factor that determines amplifier bandwidth. I suspect that the inductive reactance of the emitter feedback resistors is significant at VHF/UHF. Perhaps a microstripline layout and a change to surface mount resistors and capacitors would result in a greater bandwidth.

Band Pass Filters

The next stage in the transmit chain is the RF band pass filter unit. As this will be a multiband transceiver, I will have to build a separate filter for each band. The new BPF will use four resonators (N=4). The filters will be switched by miniature relays. This is quite expensive because two separate relays are used for each band. The miniature Toko type inductors of the old rig have been replaced by home-made inductors wound on T50 powdered iron cores. The new filter unit should be more suitable for use in a high dynamic range receiver.

The filter design is based on the same formula as the 144MHz BPF from September 2010 *RadCom*. The design pro-

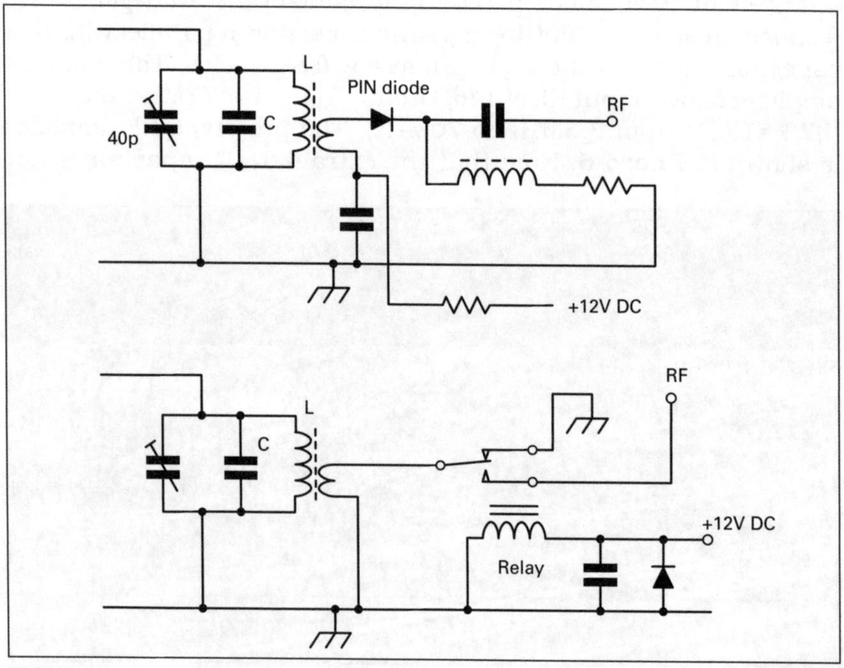

FIGURE 15: Link coupling can be used with PIN diode or relay switching.

cedure is as described in *The Handbook of Filter Synthesis* by A Zverev and in several publications by Wes Hayward, W7ZOI.

I decided to use inductive link coupling at the filter input/output because this offers the greatest flexibility in the choice of switching method. **Figure 15** shows how this arrangement is equally well suited to relay or PIN diode switching. So far, I have built and tested five sets of filters covering the 160, 80, 40, 17 and 10m bands. The filter configuration is shown in **Figure 16**. The filter is a 4th order Butterworth type designed for 50Ω I/O impedance. Working through the design equations for the 17m band (18.1MHz) gives the following component values. The inductors are 15 turns on a T50-6 (yellow) core with an inductance of 0.9μH. An unloaded Q of 200 is assumed for each resonator. This is very close to the measured value of Qu. Each resonator is tuned to resonance by a 40pF trimmer capacitor. A slightly wide filter bandwidth of 850kHz was chosen because it gives reasonable values for the coupling capacitors and low insertion loss of about 2dB.

C0=85.9pF
C12/C34=3.3pF
C23=2.2pF
C1/C4=82.6pF (47pF + trimmer)
C2/C3=80.4pF (47pF + trimmer)
Rp=1816Ω

The turns ratio for the I/O coupling coils is $\sqrt{(Rp/50)}$=6. This calls for a rather inconvenient 2.5 turns. The filter was tested with three turns and two turns. As two turns gave the best passband shape and I/O return loss, this value was used in the finished design. The completed filter is shown in **Photo 7**. A QUCS simulation showing

PHOTO 7: The finished 17m filter.

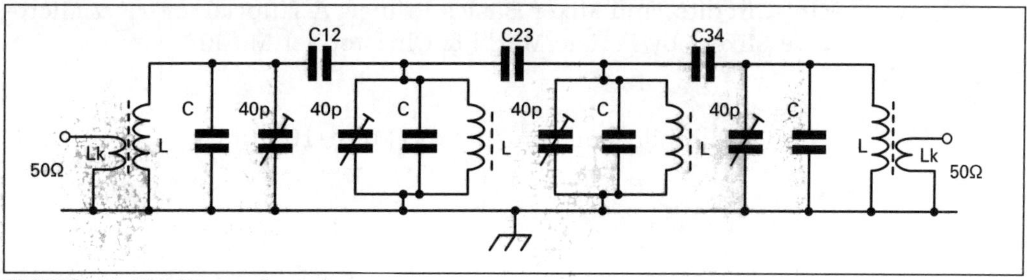

FIGURE 16: 17m filter configuration.

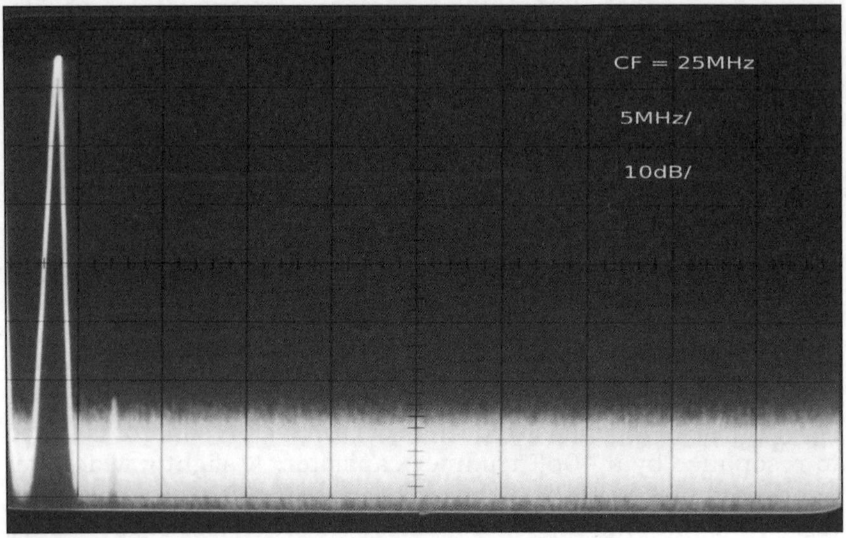

PHOTO 8: The output spectrum is remarkably clean, as shown in the 80m spectrum plot here.

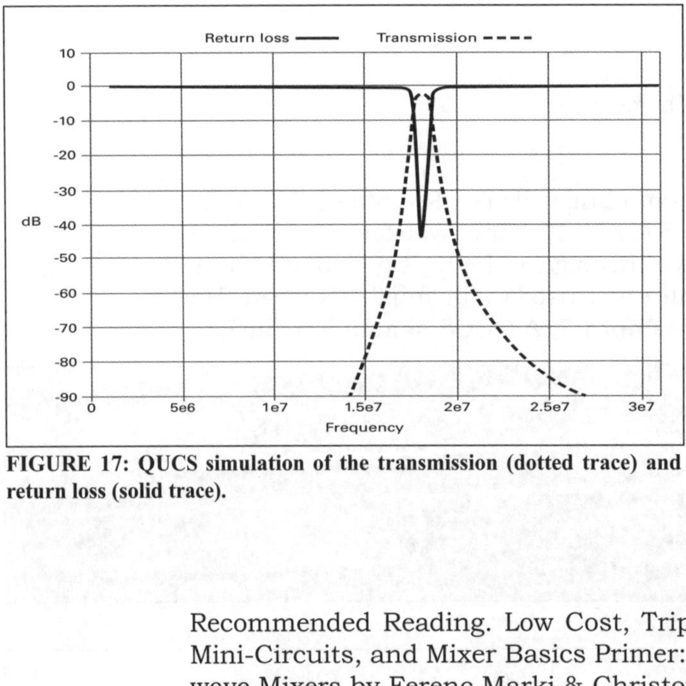

FIGURE 17: QUCS simulation of the transmission (dotted trace) and return loss (solid trace).

transmission and return loss is shown in **Figure 17**. Measurements of the finished filter show that the QUCS model is quite accurate. The output of the circuit is quite impressive. **Photo 8** shows the spectrum of the 80m version – no harmonics or spurs are detectable.

Recommended Reading. Low Cost, Triple Balanced, LTCC Mixer, Mini-Circuits, and Mixer Basics Primer: A Tutorial for RF & Microwave Mixers by Ferenc Marki & Christopher Marki.

[2] www.fairchildsemi.com/pf/MP/MPSH10.html

3

The band pass filter unit and low-level transmit amplifier stages

BUILDING A SWITCHED band pass filter (BPF) unit with separate filters for each of the nine HF amateur bands can be quite a daunting task. If we add filters for the 5MHz and 50MHz (6m) bands, we could end up with as many as eleven filters. Using fourth order filters, that is 44 hand wound coils, 22 miniature relays and several dozen capacitors. To make the task a bit more manageable, each filter will be built as a separate sub-module. After testing and alignment, these filters will be incorporated into the main BPF unit. The constructor may choose to build a full set of filters or a smaller number of filters to cover just a few favourite bands. I have left room for at least ten filters in my BPF unit. At the time of writing, I have built and tested six sets of filters to cover the 160, 80, 40, 17, 10 and 6m bands. I will add a few more when the next batch of relays arrives. I hope you won't be put off by the apparent complexity of this project. Building the filters is reasonably easy once you have a good stock of powdered iron toroid cores, fixed capacitors and trimmer capacitors. Building the filters for the LF bands is a bit more difficult because of the larger number of turns on the inductors. The large ratio of fixed to variable capacitance in the LF resonators gives a lot less scope for tuning out construction errors at the testing/alignment stage. If you start with the LF filters, your task will get progressively easier as you work your way up through the bands. I found the 10m and 6m filters were easy to build and align.

The filter for the 17m band was described in detail in the previous chapter, further information can be found on the web **[3]**. Each filter uses Amidon T50 type powdered iron cores for the inductors. The filters from 1.8 to 10MHz use the T50-2 (red) core. Coils wound on this core will have an inductance of approximately $N^2 \times 5$nH where N is the number of turns. The filters from 10MHz to 52MHz use coils wound on T50-6 (yellow) cores. These coils have an approximate inductance of $N^2 \times 4$nH. Make sure you use the correct cores. Using random cores from the junkbox is almost sure to end in failure, even if the cores happen to be the correct colour. Toroid colour codes are not universal.

I have written a very simple computer program that can be used to calculate the component values. The C source code for this program can be found on the project web page [3]. The key lines of this program are:

```
w = 2*PI*f;
C0 = 1/((w*w)*L);
k=0.8409;
Ck_12 = C0 * ((k*B)/f);
C1 = C0-Ck_12;
k=0.5412;
Ck_23 = C0 * ((k*B)/f);
C2 = C0-Ck_12-Ck_23;
q=0.7654; Qu=200;
Qe = (q*f*Qu) / ((B*Qu)-(q*f));
Rp = w*L*Qe;
```

Note that 'w' is used in place of omega 'ω'.

Practical Work

We will start with the filter for the 160m band, the filter centre frequency of which is 1.9MHz. Most filter design calculations define the bandwidth as the frequency difference between the upper and lower -3dB frequency points. This is not particularly useful when designing a maximally flat Butterworth filter or a Chebyshev filter with passband ripple of much less than 3dB. Attenuation graphs from Zverev **[4]** show that a 4th order Butterworth filter will have a -1dB bandwidth that is 0.84 times the -3dB bandwidth. A filter with a -3dB bandwidth of 300kHz will easily cover the 160m amateur band of just under 200kHz (300kHz*0.84=252kHz).

Applying the usual rule-of-thumb that says XL should be in the range 50-150Ω, I decided to use an 8μH inductor for L1, 2, 3 and L4. This is an inductive reactance of (2*π*1.9*8) = 95.5Ω. Rearranging the earlier formula for the inductance of a coil on a T50-2 core gives us √(8000÷5) = 40 turns. 40 turns of 0.375mm enamelled copper wire (Maplin YN86T) on a T50-2 core showed a measured inductance of 8.03μH, surprisingly close to the required value. The standard formula for finding the value of capacitor required to make a resonant LC circuit is based on the basic Hz, Henry and Farad units. The modified formula $106 \div ((2 \cdot \pi \cdot f)^2 \cdot L)$ allows us to use the more familiar units of MHz, μH and pF. This can be used to find the total capacitance for each resonator (nodal capacitance).

$$106 \div ((2 \times \pi \times 1.9)^2 \times 8) = 877\text{pF}.$$

To find the correct values for the inter-resonator coupling capacitors, we can use Zverev's k (coupling) values as used in our previous examples:

k12 = 0.8409 and k23 = 0.5412.
877 * (0.8409*0.3)/1.9 = 116pF
877 * (0.5412*0.3)/1.9 = 75pF

I used 100pF in parallel with 15pF for C12. This is within 1% of the required value. 75pF is a standard E24 value so you can use one if it is available: I had to resort to using 68pF in parallel with 10pF. Subtracting the coupling capacitor values from C0 gives the correct values for C1, C2, C3 and C4. 877-115 = 762pF for C1 and C4 and 877-115-78 = 684pF for C2 and C3.

End Q (Qe) and optimum load resistance (Rp) is:
Qe = (0.7654*1.9*200) / ((0.3*200)- (0.7654*1.9))
= 4.968
Rp = ω*L*Qe = 475Ω

A Qu value of 200 was used for all filters except for the 6m filter where the measured value of 89 was used. The usual model of an LC resonant circuit shows loss resistance as a resistor in series with the inductor. Qu is simply XL/r. Capacitors are often assumed to be completely lossless. This approach was probably valid back in the days when air spaced capacitors with large metal plates had Qu values of over 1000. I have learned to my cost that some modern miniature ceramic capacitors can be quite lossy.

Now that we know the value of all components and the optimum I/O load resistance, all we need to do is add the I/O coupling windings to L1/L4. The required turns ratio is √(Rp/50) = √(475/50) = 3(ish). For our 40 turn coil: 40/3 = 13 turns.

Figure 18 shows the filter schematic.

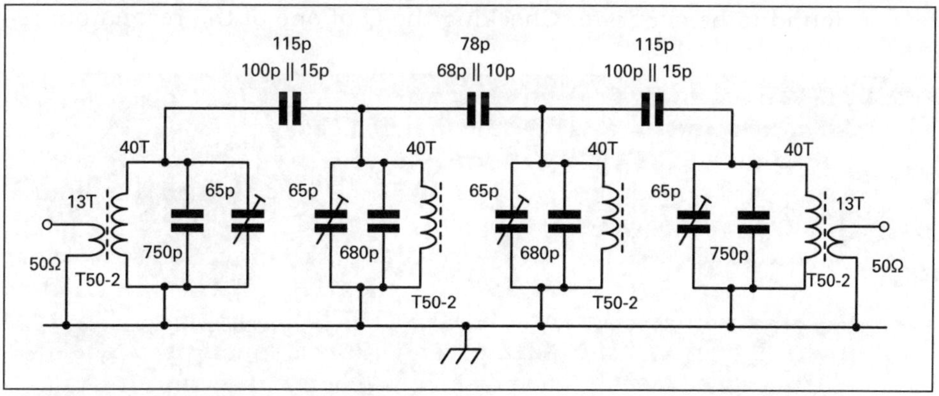

FIGURE 18: Circuit diagram of the 160m filter. Other bands use the same circuit with different component values.

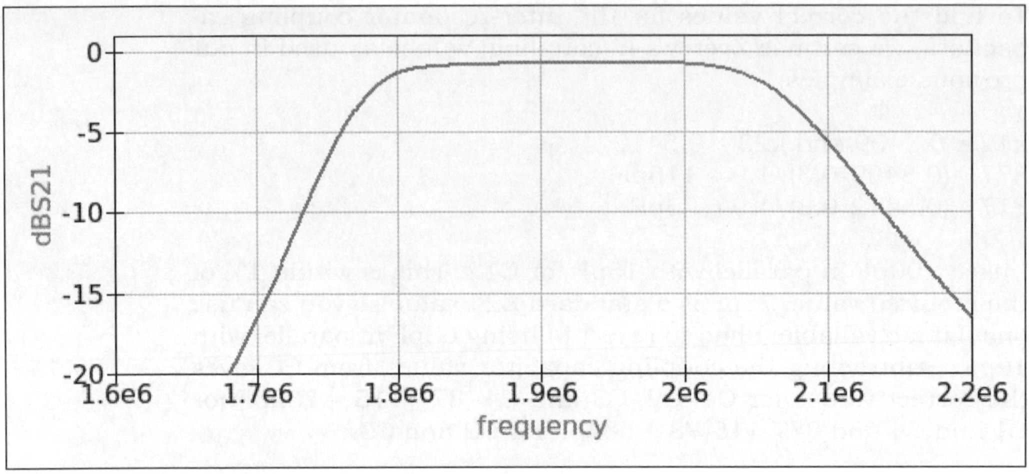

FIGURE 19: Expected response for a resonator unloaded Q of 200.

Figure 19 shows the expected response for a resonator Qu value of 200. Insertion loss of below 1dB is unlikely to be achieved in practice. Passband ripple of well below 1dB should be readily achievable. The QUCS simulation shows the expected -3dB bandwidth of 300kHz and -1dB bandwidth of around 250kHz. The filter was built on a 135x26mm strip of double sided PCB laminate. The finished unit is shown in **Photo 9**. The filter is built in the middle of the board using point-to-point wiring. The switching relays can be seen at each end of the board. I used 13 turns of insulated wire for I/O coupling. The wire was stripped from four core telephone cable. Instead of using trimmer capacitors, I used hand picked fixed capacitors for the 160m filter. The capacitor values were measured using the VK3BHR LC meter. Carefully selected 680pF capacitors were used for C2/C3. I managed to make a couple of 762pF capacitors from selected parallel pairs of 560 and 220pF. The first test of the filter gave very disappointing results. The passband was accurately centred on 1.9MHz but insertion loss was around 6dB and return loss was all over the place! All component values were checked and found to be accurate. Checking the Q of one of the resonators re-

PHOTO 9: Finished 160m filter. Filters for other bands appear similar.

vealed that Qu was just 26! The capacitors were miniature ceramic disc types that came from a very low cost kit. Substituting a high quality air spaced capacitor resulted in a radically improved Q of just over 250. Clearly these cheap and nasty capacitors are just not up to the job. When the capacitors were replaced by larger ceramic types, filter insertion loss improved to an acceptable level of 3dB and return loss is now better than 20dB (SWR 1.2:1) across the entire 160m band.

Construction

The remaining filters were built on identical 135x26mm strips of double sided PCB. The filter specification, component values and toroid details are listed in **Table 1**. Six of the filters have been built and tested. The untested filters were designed using the same formula. Any amendments and additional filter details will be posted on the web [3]. Most of the filters use a combination of fixed and trimmer capacitors. 65pF trimmers were used for the 80m filter. Filters for the higher bands used cheap and readily available 40pF trimmers that can be bought on eBay for about 25p each. The T50 toroids were bought in bulk from a well known US seller **[5]**. None of the other filters were as troublesome as the 160m filter. The 80m filter performs particularly well, with a measured insertion loss of 1.5dB, passband ripple well under 1dB from 3.5 to 3.8MHz and I/O return loss of 20-30dB.

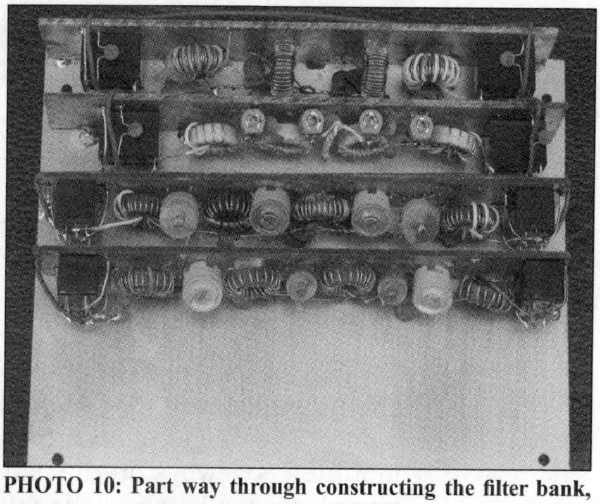

PHOTO 10: Part way through constructing the filter bank, with four sets completed.

Each filter is mounted vertically on a sheet of PCB laminate. **Photo 10** shows the first four filters in the BPF unit. The filters are soldered to the main PCB. The first filter is seam soldered along the entire length of the back of the filter PCB. The component side is soldered to the main ground plane at several strategic points close to C1/2/3/4 and the I/O switching relays. There is less working room for filters at the centre of the cluster, but you

Band	-3dB BW	L	C0	Ck12	Ck23	C1/C4	C2/C3	Rp	Toroid	Built/Tested
Generic filter: -3dB BW = f/10. Normalised to 1MHz.										
1MHz	100kHz	10μH	2533pF	213pF	137pF	2320pF	2183pF	500Ω	45/14 turns T50-2	No
160m	300kHz	8μH	877pF	115pF (100+15)	78pF (68+10)	762pF (750+trim)	684pF (680)	475Ω	40/13 turns T50-2	Yes
80m	400kHz	4.5μH	422.5pF	39pF	27pF	383pF (330+trim)	356pF (330+trim)	747Ω	30/8 turns T50-2	Yes
40m	400kHz	2.2μH	228.4pF	10pF	6.8pF	218.4pF (180+trim)	211.6pF (180+trim)	1430Ω	21/4 turns T50-2	Yes
30m	300kHz	1.4μH	177.4pF	4.7pF	3.3pF	172.7pF (150+trim)	169.4pF (150+trim) 2	628Ω	19/3 turns T50-6	No
20m	900kHz	1.0μH	126.1pF	6.8pF	4.7pF	119.3pF (100+trim)	114.6pF (100+trim)	1142Ω	16/3 turns T50-6	No
17m	850kHz	0.9μH	85.9pF	3.3pF	2.4pF	82.6pF (47+trim)	80.2pF (47+trim)	1816Ω	15/2 turns T50-6	Yes
15m	1000kHz	0.68μH	82.9pF	3.3pF	2.2pF	79.6pF (68+trim)	77.4pF (68+trim)	1600Ω	13/2 turns T50-6	No
12m	1000kHz	0.576μH	70.7pF	2.4pF	1.5pF	68.3pF (56+trim)	66.8pF (47+trim)	1900Ω	12/2 turns T50-6	No
10m	3000kHz	0.576μH	53pF	4.7pF	3.3pF	48.3pF (33+trim)	45pF (33+trim)	800Ω	12/3 turns T50-6	Yes
6m	4000kHz	0.256μH	39.3pF	2.7pF	1.8pF	36.6pF (22+trim)	34.8pF (22+trim)	870Ω	8/2 turns T50-6 Qu=89	Yes

TABLE 1: Component values for the different band filters.

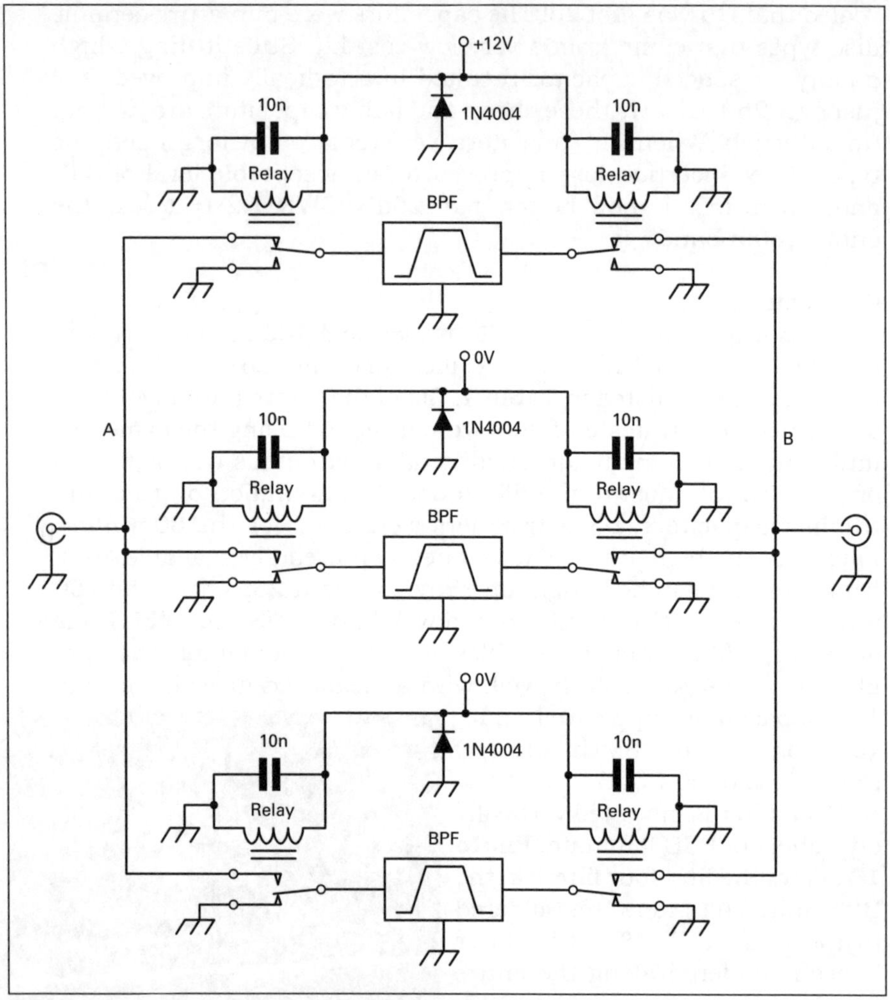

FIGURE 20: How the switching relays are arranged.

should make sure that each filter is soldered to ground at several points. The height of each filter is about 18-20mm. When they are mounted edgeways on the main filter unit, you should allow a few mm clearance between the relay pins and the adjacent filter PCB. This means you need about 23-25mm per filter or, if you prefer, one filter per inch. A small blob of adhesive from a hot glue gun was used to secure L2 and L3 of each filter to the PCB.

Figure 20 shows the arrangement of the switching relays. I used Rayex LU series subminiature relays (Maplin N15AW) with a 12VDC coil. Any similar miniature SPDT or DPDT relay should be equally suitable. The relays are wired so that the I/O terminals of all unused filters are shorted to ground. This gives much improved isolation at virtually no extra cost. There is no requirement for the filters to be mounted in any particular order, eg 160, 80, 40... There may even be a case for mounting the filters in a staggered manner so that filters for adjacent or harmonically related bands are physically separated, for example 160, 17, 40, 12, 80... The 10nF capacitor across each relay coil prevents RF leakage around the filter via capacitive coupling between the relay contacts and coil. Isolation can be further improved by placing a ferrite bead on the 12VDC wire to each relay. The 1N4004 diodes suppress negative voltage transients caused by back-EMF from the relay coils. The choice of diode is not critical. Other 1A rectifier diodes like the 1N4005/6/7 would be equally suitable.

The wires connecting the switching relays to the 50Ω input/output connectors should be as short and straight as possible. These wires are marked A and B in the schematic. Filters for the highest frequency bands (10m and 6m) should be placed near the centre of the filter cluster so that the length of the connecting wires is as short as possible. A extra few cm makes no difference on the lower frequency bands where wavelength is measured in several tens of metres

Testing

I used a spectrum analyser with tracking generator for most of the filter tests. The home made return-loss bridge from Homebrew January 2009 was used for return loss measurements. Filter insertion loss was generally between 1.5-3dB for the six filters built so far. Measured I/O return loss was generally around 20-30dB, which is in the expected range for this type of filter. The 17m filter return loss was a bit lower at just 16dB (SWR=1.4:1). I didn't bother to investigate the reason for this because I was quite satisfied with the IL, BW and passband ripple performance of this filter. Filter on/off isolation was generally better than 70-80dB. This represents the limits of my test instruments rather than the ultimate attenuation of the switching relays. Stopband attenuation is generally greater than 60dB. This type of filter has an asymmetrical response that gives greater attenuation of frequencies below the passband. This works well in practice because the transceiver will have band switched low pass filters which will greatly reduce the level of any spurious signals at frequencies well above the BPF passband.

Low Level Transmit Amplifier

I haven't yet chosen a driver and PA for the transceiver, I will proceed with a working assumption that the PA will produce 10W and the driver will produce 1W. Assuming 10dB of gain for each stage, 100mW of drive from the low level amplifier should be sufficient. The previous chapter saw the MPSH10 chosen as the standard low power VHF transistor at EI9GQ. I have been looking for a suitable VHF/UHF power transistor in the 1W range. The surface mount Philips BFG135 looks like a possible candidate. As I have recent-lybought a new batch of 2N5109 transistors, I will use one in the prototype low-level amplifier. The 2N5109 is a 1W transistor with an fT of 1.2GHz and a proven record of excellent performance in linear RF amplifiers.

The first stage of the low level amplifier (LLA) is a parallel pair of MPSH10 transistors. This is very similar to the post-mixer amp described on page 15. The second stage is a 2N5109 common-emitter amplifier with shunt feedback and emitter degeneration. Both stages are designed for 50Ω I/O impedance. The amplifier schematic is shown in **Figure 21**.

Amplifier Construction

The amplifier was built dead bug style on a strip of PCB laminate. T1 is 10 turns of enamelled copper wire, bifilar wound on a FT37-43 or similar medium to high permeability ferrite toroid. Wire diameter is not that critical: I used 0.375mm. The second stage requires a heatsink; the 2N5109 dissipates about 500mW in this circuit. After one hour of testing, the small clip-on heatsink seen in **Photo 11** was quite hot. A larger heatsink would be desirable, especially if you are a rag chewer or use SSTV or data modes. The choke in the collector circuit is 8 turns of insulated wire on a FT37-43 toroid. As usual, the wire came from standard telephone cable. At pennies per metre, this type of cable is an excellent source of general purpose insulated hookup wire.

Testing the Amplifier

This was one of those happy projects where the initial design survived the transition from computer screen to the finished hardware without needing any modification. The measured gain is very close to the design value of 35dB. **Photo 12** shows the measured gain

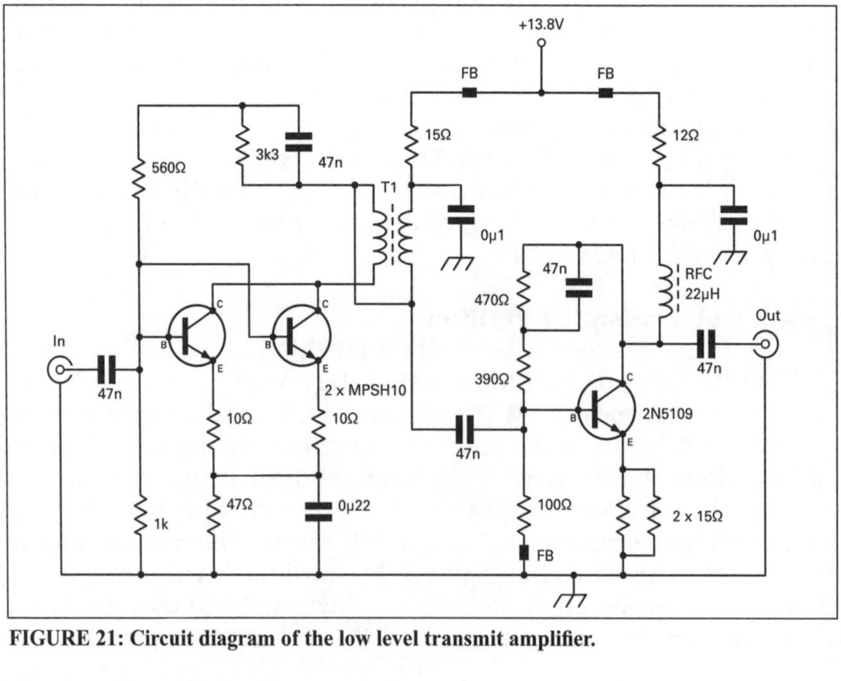

FIGURE 21: Circuit diagram of the low level transmit amplifier.

PHOTO 11: The completed low level transmit amplifier..

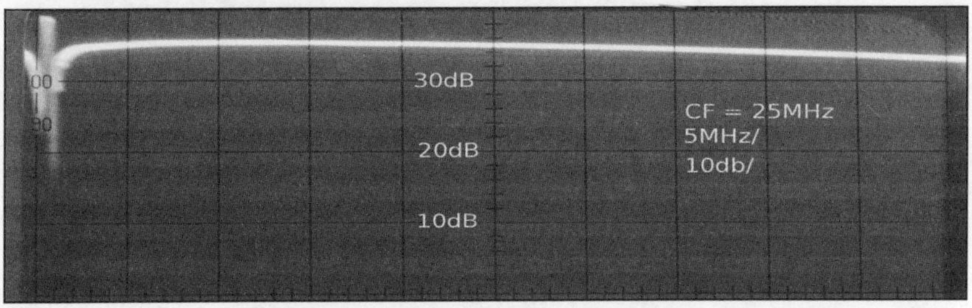

PHOTO 12: Low level transmit amplifier gain versus frequency.

from LF to 50MHz. Gain is almost perfectly flat from 1.8MHz to 30MHz and stays above 33dB up to 52MHz. Input return loss is better than 16dB from 1.8MHz to 100MHz, output return loss is better than 10dB (SWR=2:1) from 1.8MHz to 100MHz. The amplifier is very well behaved at output levels up to 200mW and saturated output is about 500mW.

All of the modules built so far were wired up on the bench. The local oscillator signal was provided by the DDS signal generator (Homebrew March 2010) and the output from the LLA was fed to my HF doublet via a 31MHz LPF (Homebrew October 2010). With the LO generator set to 17.803MHz (28.5-10.697) and the 10m BPF selected, I received a 53 report from Gary, EI8GQ at a distance of 8km. Not bad for an output of 100mW, especially since Gary uses a vertical aerial and I have a horizontal wire doublet. Reports on the modulation quality were very good.

[3] http://homepage.eircom.net/ ~ei9gq/bpf.html
[4] *The Handbook of Filter Synthesis,* A Zverev, Wiley, 1967
[5] www.kitsandparts.com

4

Broadband linear power amplifier

BROADBAND POWER amplifier design and construction can be one of the more difficult challenges for the home-brew radio enthusiast. High power amplifiers designed to run from a 12V supply can be particularly problematic. If we take the extreme case of a single ended 12V 100W transistor PA, the collector load impedance will be approximately $(Vcc-Vsat)^2/2Po$. Even allowing a minimum voltage swing to just 1V on negative peaks, this gives a collector load impedance of $11^2/200 = 0.6\Omega$. The input impedance at the base terminal of a bipolar transistor operating under these conditions is typically below 1Ω. Because transistor gain decreases with frequency, the resistive part of the input impedance tends to decrease with rising frequency. Power FETs have a slightly higher input impedance, but still in the single ohms range at HF and VHF. Working with such low impedances and high current levels is difficult for the designer and constructor. It is also potentially very costly when the magic smoke finds its way out of your power transistors.

For this project, I will attempt to push the boundaries by including the 6m band in the specifications. The low level amplifier described in the previous chapter has a very flat response from 1.6MHz to more than 70MHz. It will be a bit more difficult to achieve similar bandwidth from a power amplifier. As several commercial transceivers have broadband PAs that work from 1.8 to above 50MHz, at least we know that it can be done.

The maximum RF power output available at the collector (or drain in the case of a FET) of a 12V power amplifier with an inductor (RF choke) in the output circuit is $Vcc^2/2R$.

Figure 22 shows a simplified RF power amplifier. When the output load resistance is 50Ω, the maximum output power is 1.44W. Increasing the supply voltage to 13.8V gives a maximum of 1.9W. As the load impedance is fixed at the standard value of 50Ω, the only way to increase the output power beyond this level is to increase the output voltage swing. There are a number of ways of doing this. The most obvious method is to simply increase the supply voltage. Increasing the supply voltage to 50VDC would increase the maximum output power to around 25W. Many transistors also show better linearity at higher supply voltages, but you should make sure that the maximum voltage rating of the transistor is not exceeded. Peak collector voltage will approach twice the DC supply voltage when the output is properly terminated by a 50Ω load and even higher voltages are possible when there is a mismatch at the amplifier output. The other option is to use some kind of voltage transforming network between the transistor and the load. Single band amplifiers can use a resonant matching network for the voltage transformation. For a multi-band amplifier, the only practical solution is a broadband transformer. Conventional flux-coupled transformers and transmission line transformers are both widely used for this application. The transformer is usually wound on a ferrite core.

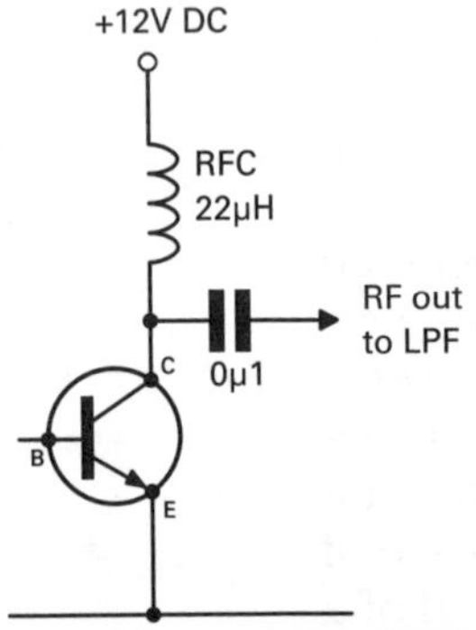

$$Vcc_{min} = \sqrt{(2 * R_L * P)}$$

$$R_L = \frac{(Vcc - V_{SAT})^2}{2 * P_O}$$

$$P_O = \frac{(Vcc - V_{SAT})^2}{2 * R_L}$$

FIGURE 22: Simple amplifier and its associated calculations (see text).

Ferrite and Powdered Iron Cores

The ferrite toroid core is one of the most mysterious components used in radio construction. To the uninitiated, a ferrite toroid has no obvious function. It is just a doughnut shaped object made from hard black stuff. However, this black stuff has some rather magical properties. Inductors used at low frequencies are often wound on a high permeability magnetic core. Transformers for mains electricity or other low frequency applications like audio transformers are usually wound on an iron alloy core. A wire with a current flowing through it is surrounded by a magnetic field. The presence of the transformer core contains and concentrates this magnetic field in a way that greatly increases the inductance of the transformer windings. The inductance increases by an amount that is proportional to the relative permeability of the core material. Permeability is usually measured relative to the vacuum permeability of free space (if such a place existed). As the difference between the permeability of air and free space is very small, foremost practical purposes, the permeability of air is also 1. The relative permeability of magnetic cores ranges from about 8 for iron powder RF toroids, to several thousand for high permeability ferrite and iron alloy cores.

A well designed power transformer is quite efficient. Most of the energy fed to the primary winding will be delivered to the load, with minimal loss. (Transformer cores are not perfectly linear devices. The alternating current supplied to the primary causes an alternating magnetic field in the transformer core. Each time the field is reversed, some energy is lost due to hysteresis in the core.) The metal cores used in mains transformers work well at low frequencies but become progressively more lossy as frequency is increased. Because iron cores are conductive, current flowing in the core will lead to eddy current losses. Such losses are reduced by using iron alloys with a relatively high resistivity and making the core from a stack of laminated sheets of iron alloy with insulation between each lamination. Powdered iron toroids for use at RF are made from tiny metal particles which are suspended in an insulating medium.

Figure 23A shows an ideal transformer. Users of electronic simulation software will be familiar with this wonderful device. It is completely lossless, works over an infinite bandwidth, the primary/secondary voltage ratio is exactly the same as the turns ratio and the impedance ratio is the square of the turns ratio. Such transformers don't exist in the real world. The transformer model in **Figure 23B** is more accurate, although this model is also incomplete. This transformer has losses due to resistance in the primary and secondary windings, capacitance of both windings and leakage inductance due to imperfect coupling between primary and secondary. Other imperfections include core losses as mentioned earlier and core saturation that causes winding inductance to vary with winding current. Non-linearity due to core saturation can be avoided by careful selection of core material and by using a core that is sufficiently large for the power level used.

Iron powder toroids like the Amidon type 2 and type 6 cores used in some of our recent projects make very good high Q inductors for use in filters. These cores have relatively low permeability. They are not prone to non-linearity caused by core saturation and have very low eddy current losses because of the way that individual iron particles are separated by insulating material. The

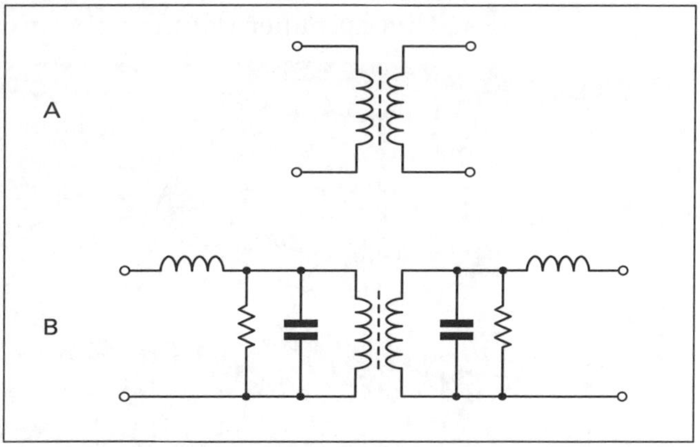

FIGURE 23: The (A) ideal and (B) slightly more realistic transformer.

low relative permeability of iron powder toroids makes them less suitable for use in broadband RF transformers. As a rough rule of thumb, transformer windings should have an inductive reactance (XL) that is several times greater than the impedance the winding is connected to. For example, a transformer winding for use in a 50Ω circuit should have an inductive reactance of several hundred ohms at the lowest operating frequency. A broadband transformer wound on a low permeability core would require a large number of turns for satisfactory operation at lower frequencies. This would result in high winding capacitance in a conventional transformer or greater phase delay in transmission line transformers.

Ferrite toroid cores are widely used in broadband circuits. Relative permeability of ferrite cores ranges from about 30 to several thousand. The two most popular types for RF applications are type 61 with μi (initial permeability) of 125 and type 43 with μi of 850. These are suitable for use in low Q broadband transformers at MF/HF and into the VHF region with increased losses. High permeability ferrites are prone to core saturation. Take care to use a core that is large enough for the application. Relatively high losses at HF/VHF make these cores unsuitable for high Q applications like filters and resonant matching networks.

Transmission line transformers wound on ferrite cores can work over a huge bandwidth. The balun transformers in some diode mixers work well from MF to UHF.

A Broadband Class A Power Amplifier

I will use a TPV3100 dual transistor **[6]** for the prototype of the PA project (**Photo 13**). Other VHF power transistors should also work well in this circuit. The use of negative feedback and Class A operation should ensure good results from a wide range of VHF transistors. I have a good stock of small FT37-43 ferrite cores, but I will have to make the broadband output transformer using only High Street components. I love a challenge!

This transceiver will be used to drive a legal limit MOSFET PA. This amplifier requires about 6-7W of drive for 400W out. A 10W PA will be a good match for the MOSFET amplifier. This is also a useful power level on the 6m band, which is not covered by my 400W linear.

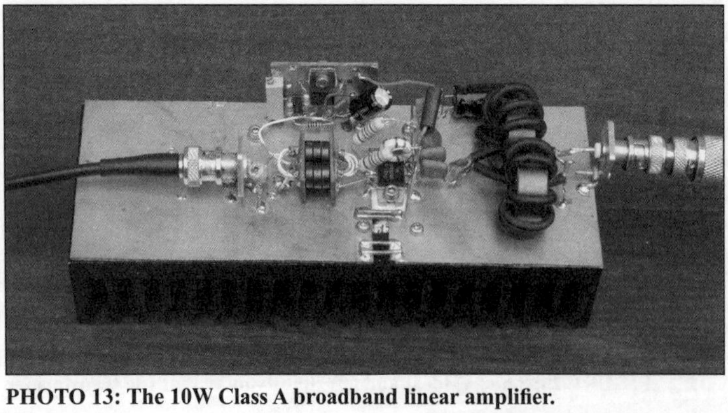

PHOTO 13: The 10W Class A broadband linear amplifier.

The TPV3100 is a dual transistor (two devices in one package) designed for linear Class A or Class AB operation at VHF. Having both devices in a single package should mean that the two transistors are well matched. Having a single mounting flange greatly simplifies the mechanical and heatsinking arrangements.

The amplifier will be in a push-pull configuration with broad-band input and output matching transformers. The datasheet figures suggest an input impedance of around 1Ω at 200MHz. As I will be using the transistors at a much lower frequency and at a lower power level, I will assume a higher input impedance of 2-3Ω for each transistor (or 4-6Ω across the push-pull inputs that are effectively in series). After a few experiments with both conventional and transmission line transformers, I decided that a conventional transformer with a turns ratio of 3:1 and impedance ratio 9:1 would give the best results. Tight coupling between primary and secondary is critically important if we are to achieve broadband operation from 1.8MHz to 52MHz. A high value of leakage inductance would limit high frequency performance. The transformer secondary winding is just a single turn made from two lengths of 4mm brass tubing (B&Q) and two end plates made from single-sided PCB laminate. I used a hacksaw to cut the copper foil on one of the end plates at a point half way between the ends of the brass tube. This forms the open end of the single turn winding. A total of six FT37-43 ferrite toroids were used for the transformer core, three on each brass tube. The primary winding is three turns of insulated wire threaded through the brass tubes. Each tube is 21mm long. **Figure 24** shows details of the transformer construction.

The measured inductance of the primary winding was 20μH. This is 226Ω of reactance at 1.8MHz. A perfect transformer would show an inductance of zero with a short circuit load. With a small strip of copper soldered across the single turn secondary, the measured primary inductance was 0.25μH, or XL=78.5Ω at 50MHz. Because of the difficulty of measuring extremely low values of inductance, the secondary winding leakage inductance was assumed to be $250nH \div 3^2$ or 27.8nH. These are just ball-park figures to be used as a starting point for choosing HF compensation capacitors. My L/C meter uses a 0.5MHz test oscillator, so the measured values could be quite far from the actual values at HF/VHF.

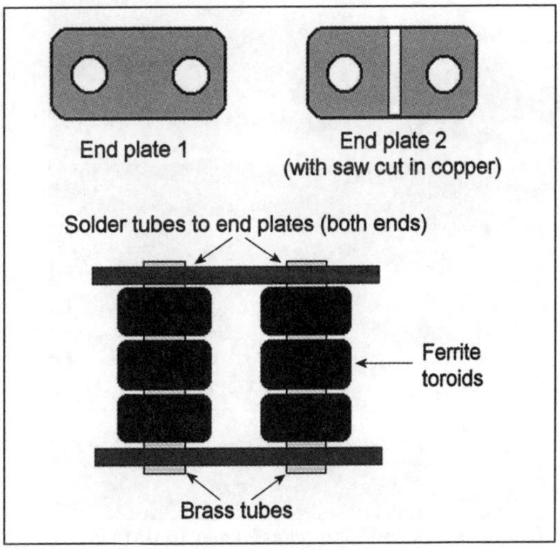

FIGURE 24: Construction of the output transformer.

The transformer secondary was terminated by a 5.5Ω resistance. I used 10Ω in parallel with 12Ω. The resistors are 0.25W metal film types. I used a tracking generator and spectrum analyser to measure return loss at the primary winding. The first test was encouraging. Return loss was

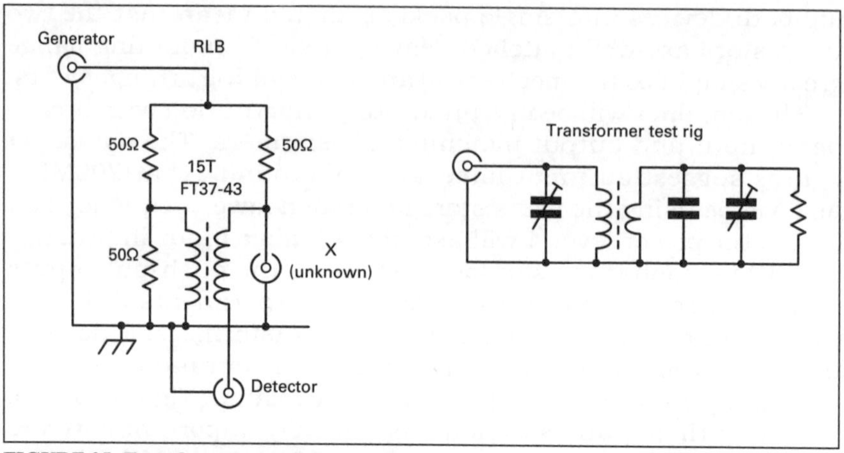

FIGURE 25: Transformer test rig (see text).

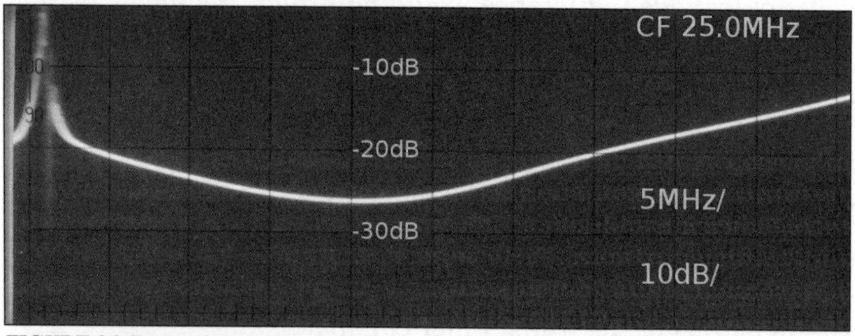

FIGURE 26: Return loss from transformer.

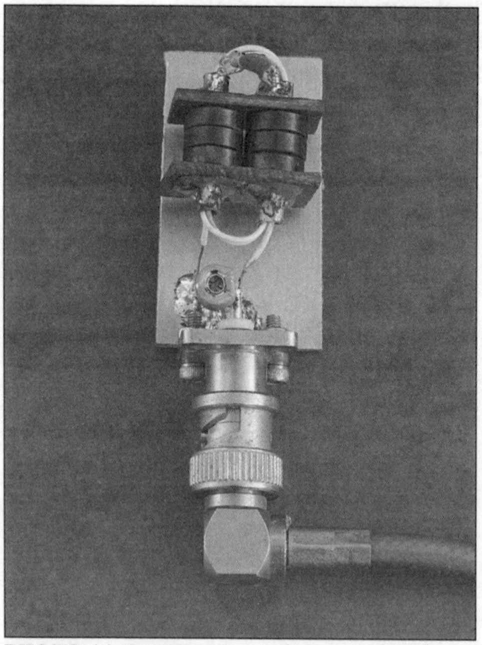

PHOTO 14: Broadband transformer in test rig.

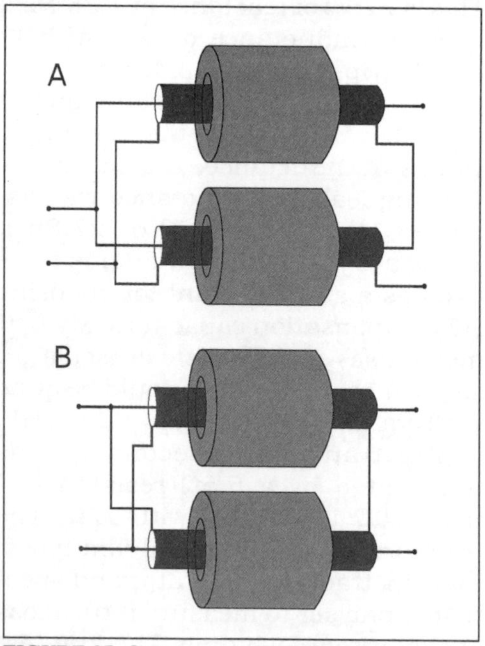

FIGURE 27: Output matching transformer schematic, showing two different configurations.

good up to about 34MHz, but quite poor around 50MHz. Placing a capacitor with XC=78.5Ω = 40.5pF across the primary and a capacitor with 9 times this value (365pF) across the low impedance secondary gave much better results. I used a 40pF trimmer that has an actual maximum C around 45pF and a 390pF fixed ceramic disc capacitor for the tests. Details of the test rig and return loss bridge are shown in **Figure 25**. A plot of return loss from LF to 50MHz is shown in **Figure 26** and the transformer and test rig are shown in **Photo 14**.

The amplifier output matching transformer is a transmission line 1:4 impedance transformer. Each transmission line is 6 turns of coax cable on a Maplin N88AB ferrite toroid. The transformer configuration is shown in **Figure 27A. Figure 27B** shows how the circuit can be slightly reconfigured to achieve the same result.

The optimum impedance for the coax is 25Ω. I could have used two parallel lengths of miniature 50Ω coax, but instead, I decided to use the nearest equivalent to 25Ω coax that I could find. Each of the two toroids is wound with six equally spaced turns of Maplin XR16 screened audio cable. This has a measured capacitance of 232pF/m and inductance of 210nH/m, indicating a characteristic impedance of $\sqrt{(L/C)}$ = 30Ω. As each coax line is just over 30cm long, cable losses are a tiny fraction of a decibel, even at 50MHz. **Photo 15** shows the completed output transformer.

The amplifier will operate in Class A. The output voltage swing of each transistor will be limited to 10V or preferably less. The output transformer presents an impedance of 50/4=12.5Ω to the transistor collectors. This is 2x6.25Ω for a push-pull amplifier. Output power will be $10^2 \div (2 \times 6.25)$ = 8W (see Figure 22) for each transistor, or 16W for the pair. In the interests of linearity, the amplifier will normally be run at a maximum output of just 10W.

PHOTO 15: Close-up of the practical output transformer.

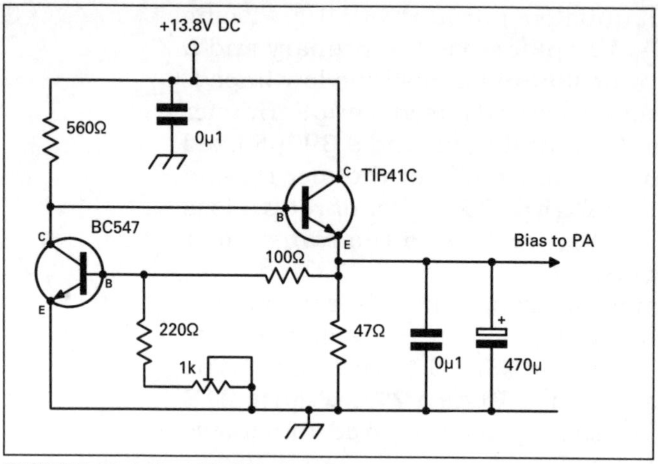

FIGURE 28: Adjustable bias circuit.

The bias circuit is shown in **Figure 28**. This circuit can generate an adjustable bias voltage from 0.75V to 1.1V.

The amplifier was built on a strip of PCB laminate on the back of a reasonably large heatsink with a thermal resistance of about 1.5°C/W. **Figure 29** shows the circuit diagram. The 56Ω resistors are 2W metal film types. The 100nF capacitors in the collector to base feedback network are critically important components. They should be good quality ceramic types with a voltage rating of 100V or more. The ferrite bead in the collector supply is a Maplin N98AB. T2 is 8 turns of 0.375mm enamelled wire on a FT50-43 core. For clarity, the schematic shows each coax line in T3 as a single turn. Each core actually has 6 turns, as shown in Photo 15. The collector supply for the transistors is shown as Vcc_1 and Vcc_2. These points are normally connected together, but they can be separated to check the current balance between the transistors. Photo 13 shows the finished amplifier.

Testing

The bias voltage was set to minimum by adjusting the 1k pot to maximum resistance. Next a 50Ω load was connected to the output, the 13.8VDC supply was connected and the bias adjusted

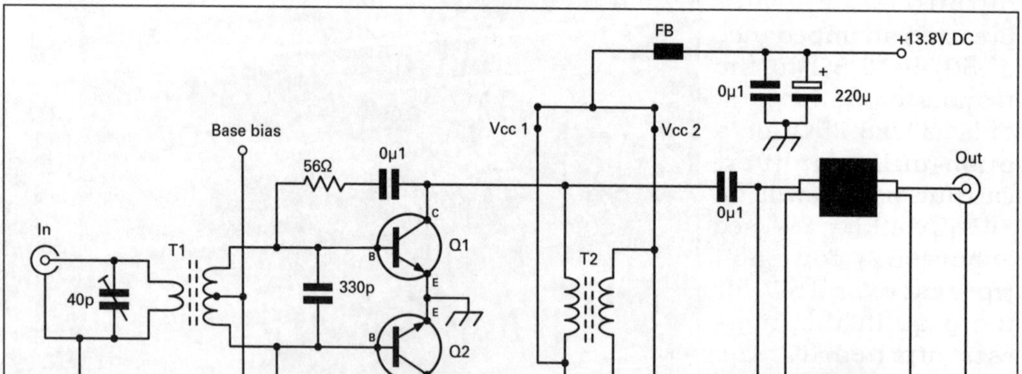

FIGURE 29: Wideband linear amplifier circuit.

for a standing current of 2.5A (a DC input of 34.5W). After one hour of running at this level, the transistor header was at 65°C and the heatsink was warm, but not excessively hot. I reduced the value of the 390pF capacitor across the secondary of T1 to compensate for the transistor base capacitance. A value of 330pF gave best results. The 40pF trimmer was adjusted for peak output at 50MHz. **Figure 30** shows a plot of the amplifier frequency response from LF to 100MHz. Peaking the T1 frequency compensation for a peak at 50MHz caused a slight 'glitch' at 38-40MHz, but as this is far removed from any amateur bands, it is of no consequence.

Gain is very flat up to 34MHz and still within 1dB of the MF/HF value at more than 50MHz. Power output for the test was 10W. At this point, I decided to have another little tweak at the T1 trimmer capacitor. The trimmer is new and quite stiff, so I have to push the screwdriver quite hard to turn it. During this operation, I heard a slight click. This was followed by a smell of burning components! The over-current trip shut down the power supply. The post-mortem examination revealed that the TIP41 and the BC547 in the bias circuit were both blown. I decided that the most likely cause of the failure was a short circuit between the T1 secondary and ground while I was tweaking. The TIP41 was short circuit from C-B-E, so when I removed the pressure from the screwdriver, the full 13.8V supply would be applied to the base of the two RF transistors. As it turned out, the fault was actually caused by a short circuit between the primary and secondary of T1, probably caused by the heat of the soldering iron during the test rig experiments. Replacing the primary of T1 with three turns of nice new insulated wire and changing the two transistors in the bias circuit cured the problem. The TPV3100 was not damaged.

The bias was increased to 3A (or 41W) for on-air testing. A few transmissions on 160m resulted in some very good reports on the audio quality. I was surprised to hear my own 10W 160m signals quite clearly on the HB9FX Web-SDR receiver in Switzerland.

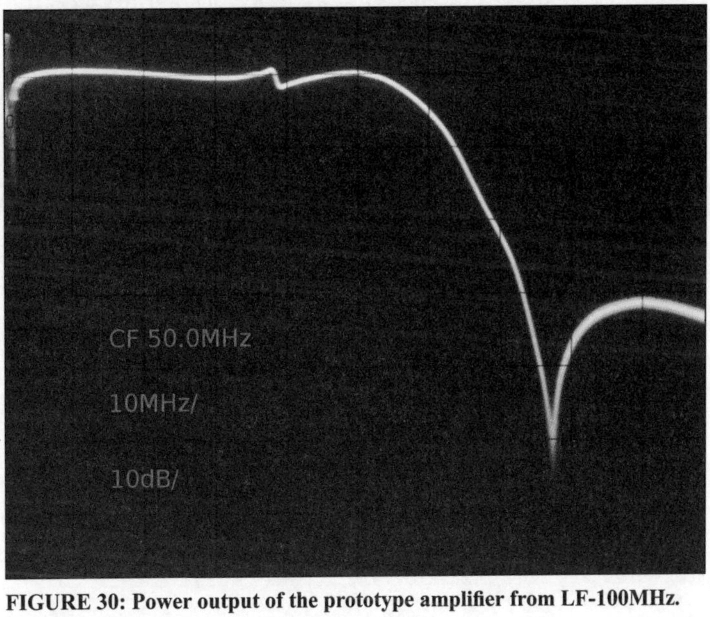

FIGURE 30: Power output of the prototype amplifier from LF-100MHz.

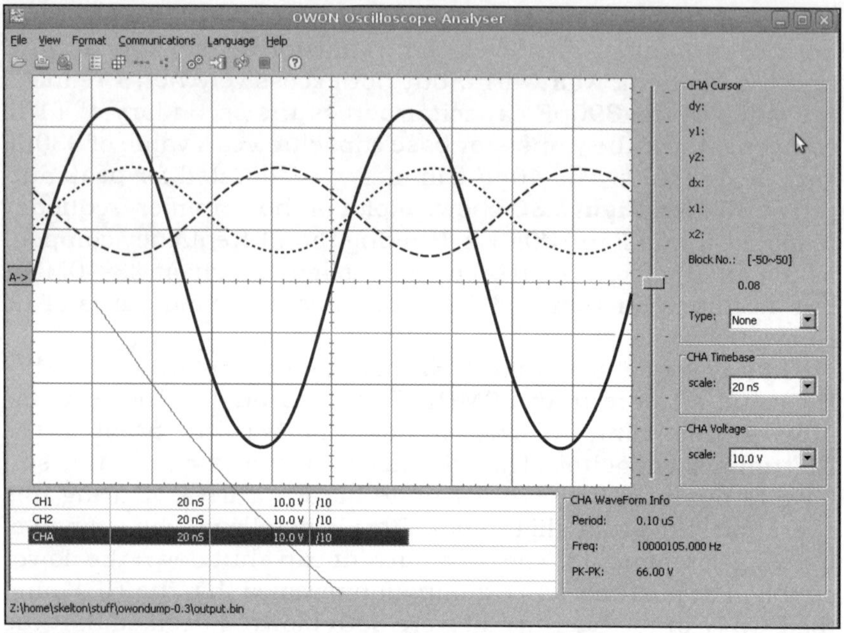

FIGURE 31: Measured collector voltages *(dotted & dashed lines)* and output voltage *(solid line)* at 11W output.

Figure 31 shows the voltage measured at the collector of both transistors (dotted and dashed lines) and the output voltage (solid line). Output for this test was 66Vpp into 50Ω, or just under 11W. Note that no LPF was used for this test. Because the amplifier is well balanced, second harmonic suppression is excellent, at -40 to -50dBc in the 5-10W output range. Third harmonic suppression is not quite as good, but still reasonable at -30 to -35dBc.

[6] Motorola TPV3100 data sheet.

5

Power amplifiers and the low pass filter unit

THE CLASS A power amplifier described in the previous chapter uses a rare and exotic VHF dual transistor. It should be possible to substitute a well-matched pair of HF or VHF power transistors. Suitable devices will have a maximum dissipation rating of at least 50W each and fT of several hundred MHz. CB PA transistors like the 2SC1969, 2SC1307, 2SC2312 and 2SC1945 should work at frequencies up to 30MHz, but they will probably run out of gain by 50MHz. These transistors are not rated for Class A operation, so you should reduce the standing current to about 250mA for each transistor, biasing the amplifier for Class AB operation.

The first project **(Photo 16)** is a broadband power amplifier based on a pair of Mitsubishi RD16HHF1 MOSFETs. These devices are used in the final stage of some of the more modern CB rigs. Unlike the power MOSFETS we have used in previous projects, they are designed to work with a 12-14V DC supply.

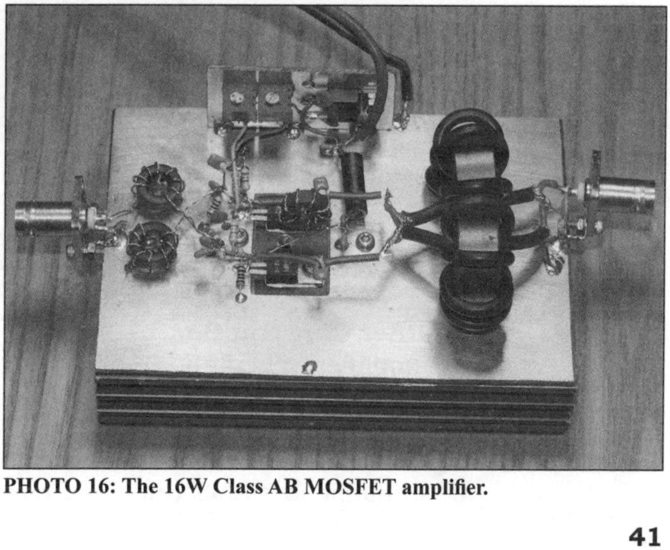
PHOTO 16: The 16W Class AB MOSFET amplifier.

The RD16HHF1 is specified for HF RF power amplifier applications. The data sheet test circuit is for a 30MHz amplifier. Input impedance is specified as 20.02-j89.42 at 30MHz, 20W out, Vdd=12.5V. S21 (forward gain) parameters

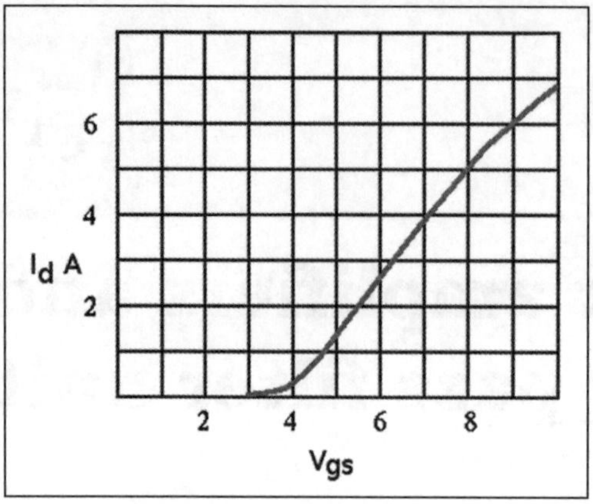

FIGURE 32: Vgs-Ids (gate voltage/drain current) curve for the RD16HHF1.

from the data sheet suggest that the device should have useful gain at frequencies up to the low to mid VHF region.

The amplifier output matching transformer (T3) has an impedance ratio of 1:4. When the output is terminated by a 50Ω load, the drain-to-drain load impedance is 12.5Ω, or 6.25Ω for each FET. With a 13.8V DC supply, the theoretical maximum output is $Vdd^2/2R$ = $13.8^2/12.8$=15.2W per transistor, or just over 30W for the pushpull pair. As with the bipolar transistor design in the previous chapter, I will limit the drain voltage swing to an absolute maximum of ±10V. This allows a maximum output of $10^2/12.8$ = 8W *2 = 16W. The amplifier will be biased for Class AB operation. Standing current will be set at a relatively high 500mA per device, or 1A total. This keeps the gate voltage within the most linear part of the Vgs-Ids (gate voltage/drain current) curve, which typically falls between +4.7V and +7V. See **Figure 32.**

The input resistance and reactance of a MOSFET tends to vary with drive level and signal frequency. It is common practice to place a resistor of relatively low value in parallel with the input of MOSFET amplifiers. The resistor provides a constant load under all conditions and because of its low value; it tends to swamp the input resistance/reactance of the FET. This gives a better input match over a wide bandwidth. This arrangement leads to a loss of amplifier gain due to power dissipated as heat in the resistor. In the case where the resistor value is lower than the FET input impedance, more than half of the input power is lost and amplifier gain will be reduced by more than 3dB. As RF power FETs typically have MF/HF gain of more than 20dB, a gain loss of 3-6dB may be considered an acceptable tradeoff for greater bandwidth and improved input matching.

The input matching transformer (T1) is another 4:1 transmission line type. It is a smaller version of the output transformer, with the input/output connections reversed so that it presents an input impedance of 50Ω when the output is terminated in 12.5Ω. The FET input resistance of 20-odd ohms is shunted by a 10Ω resistor. This gives a resistance of about 6.66Ω (or 13.33Ω for the push-pull pair of devices). The 89Ω capacitive reactance (about 60pF at 30MHz) is effectively swamped by the low value of gate circuit resistance.

Amplifier Construction

The amplifier schematic is shown in **Figure 33.** As usual with RF power amplifiers, most of the builders' effort is applied to the mechanical construction and thermal design. As it is just a single stage amplifier, the electrical circuit is very simple. To ensure that the amplifier is stable under all conditions and has maximum

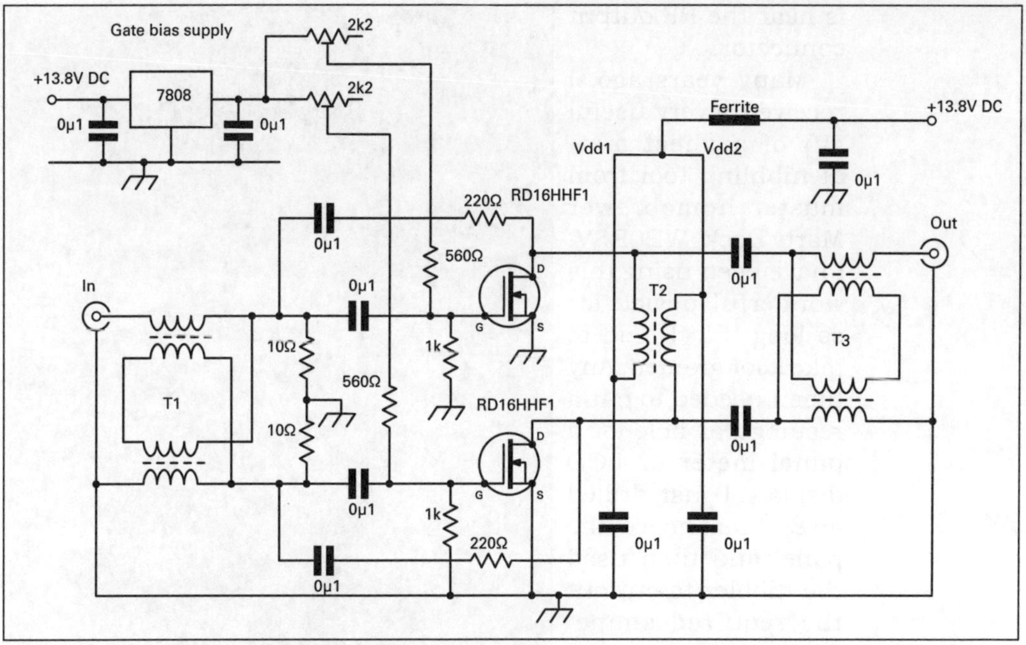

FIGURE 33: The Class AB RF power amplifier circuit diagram.

bandwidth, the builder should follow best HF/VHF practice. All leads should be as short and straight as possible. Decoupling capacitors should be soldered directly to the copper ground foil using short lead length of no more than a few mm. To ensure good circuit balance, the amplifier should have a physically symmetrical layout and the FETs should be well matched. At the very least, they should come from the same packing tube and have identical date and gain group codes.

The amplifier was built on a strip of PCB laminate. The heatsink was drilled and tapped using a 2.5mm drill and M3 tap to make the mounting holes for the PCB and transistors. **Figure 34** shows the mounting details. So far, I have drilled and tapped six holes, four for the PCB mounting screws and two for the MOSFET heatsink tabs. The PCB mounting screws are strategically placed along the centre line between the two transistors. Hole 1 is close to the RF input connector, 2 is within a few mm of the source leads, 3 is close to the drain supply decoupling capacitors and 4

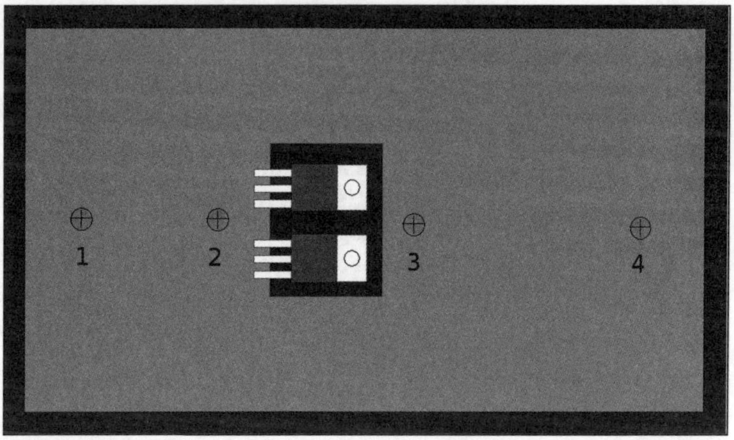

FIGURE 34: Mounting details for the amplifier board on the heatsink (see text).

is near the RF output connector.

PHOTO 17: How I used the new nibbler to cut a rectangular hole in the PCB so that the MOSFETS can be screwed to the heatsink.

Many years ago, I received a very useful gift of a sheet metal nibbling tool from master homebrewer Marty Beck, WB0ESV. I have been using this wonderful device for so long that I tend to take it for granted. Any time I needed to cut a rectangular hole for a panel meter or LCD display, I just drilled an 8-10mm hole in the panel and then used the nibbler to cut out the required shape. After twenty years of abuse, the nibbler has finally broken! Local tool shops don't stock this type of tool, so I had to go to an American eBay seller to get a replacement. **Photo 17** shows how the new nibbler was used to cut a rectangular hole in the PCB so that the MOSFETS can be screwed to the heatsink. The input transformer T1 is made from two twisted pair transmission lines. Each line is 8 turns of 0.375mm enamelled copper wire, bifilar wound on a FT50-43 toroid core. T2 is 8 turns of 0.375mm twisted pair, bifilar wound on another FT50-43. The output matching transformer T3 is identical to the one in the previous amplifier project. It consists of two HEM3011 (Maplin N88AB) toroid cores, each wound with six turns of Maplin XR16 screened cable. The long ferrite sleeve on the drain supply lead is a HEM3021 (Maplin N98AB). All capacitors are ceramic types. Use good quality 100V (or more) capacitors for the negative feedback network and the drain decoupling capacitors. I used 0.6W rated 220Ω metal film resistors for the drain to gate feedback network. These get quite hot at full output. 2W metal film resistors should be used instead. The gate bias circuit is very simple. A 7808 regulator provides a stabilised 8V DC bias supply. The 2.2k bias adjustment pots form part of a simple potential divider, which includes the 560Ω and 1k resistor in the MOSFET gate circuit. One nice feature of the RD16HHF1 is that the drain terminal is not connected to the heatsink tab. This means that the FETs can be mounted directly on the heatsink without using an insulated mounting kit. The heatsink tab is connected to the source terminal, which is also connected to ground. Note that the RD16HHF1 pinout is different from most other TO220 packaged transistors. The connections are G-S-D and not the more common G-D-S configuration. See the data sheet for full details.

Testing
The amplifier output should be terminated by a 50Ω dummy load. If possible, use a current limited 13.8V supply for the tests. If a current limited supply is not available, use a 2A fuse or a small value resistor (around 2Ω) in series with the positive DC supply lead. After checking all connections one last time, set the bias pots for minimum bias voltage (maximum resistance) and connect the DC supply to the amp. The current drawn from the supply should be close to zero. Adjust one of the bias pots for 250mA, then adjust the other until the total current draw is 500mA. Check the temperature of the transistor packages (they should be cool) to make sure they have a low thermal resistance to the heatsink. If all seems well, apply some RF to the input and check the output level using a power meter or oscilloscope. You should see a power gain of around 10dB. For high power tests, set the power supply current limit at several amps and/or remove the protective resistor or fuse from the supply lead. For final testing, the bias should be set at 500mA per FET, giving a total standing current of 1A. I used the tracking generator of my spectrum analyser as a signal source. The low level amplifier (Homebrew February 2011) was used to bring the tracking generator signal up to 200-300mW. This limits my initial power and bandwidth testing to just 2-3W output. **Photo 18** shows the measured gain at an output power of 2W.

Gain is remarkably flat at around 10dB from 1.5MHz to 60MHz and actually increases to about 12dB in the 70-90MHz region before rolling off at 100MHz. This is by far the best results I have seen from a HF broadband PA. It seems these Mitsubishi MOSFETs are a lot better than the old generation of bipolar CB PA transistors. This PA should perform well from 160m-6m. The measurements suggest that 4m is also a distinct possibility at some time in the future...

As the current configuration of the HF transceiver project doesn't have enough power output to drive this new PA, I used the driver stage of my old homebrew rig for the high power tests. This 2SC2166

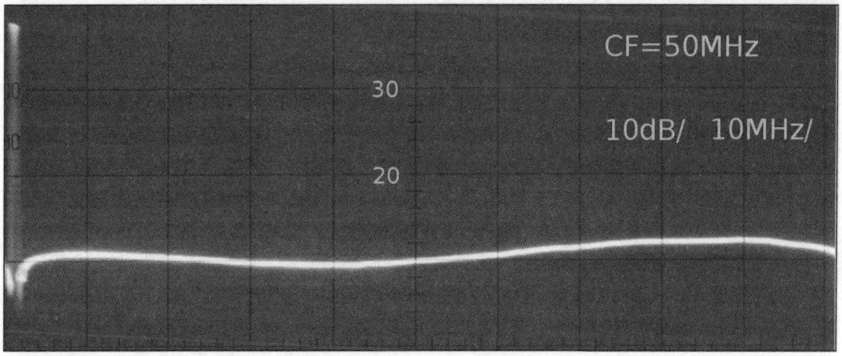

PHOTO 18: Measured gain of the amplifier at an output power of 2W..

stage provided the necessary 1-2W to drive the new PA. My first test on 80m resulted in a 59+ report from Tony, GM7KFS in Argyll. Reports on the signal quality were very good. I used an output power of 10W for the on-air tests. Saturated output power into a dummy load was measured at just under 25W.

Low Pass Filter

The next stage in the transceiver project is the low pass filter (LPF) unit. As both of the PA units built so far are balanced push-pull circuits, even order harmonics (2nd, 4th etc) are quite well suppressed. In the case of the Class A amplifier, the second harmonic is already more than 40dB below the wanted output. A simple 5th order LPF would probably provide adequate harmonic suppression. As the extra cost of building a 7th order filter is minimal, I will build a set of 7th order filters (7 reactive elements in each filter). There are a few options available: I could choose to build a Butterworth type of filter that would probably give adequate harmonic suppression or a Chebyshev type that will give greater harmonic suppression at a cost of slightly greater passband ripple and higher SWR at the filter input. A Chebyshev type filter with a very low passband ripple of 0.1dB or less is probably the best choice. A 7th order 0.1dB Chebyshev will have a maximum input/output SWR of 1.36:1, attenuation of 57.7dB at 2fc and 84.8dB at 3fc.

PI (parallel capacitor input) or T (series inductor input) low pass filters are both equally suitable for the job. A 7th order PI LPF will have three inductors and four capacitors. The T equivalent would have four inductors and three capacitors. As capacitors are generally smaller, lighter and cheaper than inductors, the PI type is more widely used. **Figure 35** shows a 7th order PI LPF. Normalised component values are shown in **Table 2** for a 1MHz filter with I/O impedance of 50Ω. To calculate suitable values for any frequency greater than 1MHz, simply divide the normalised value by the required filter cutoff frequency. If you want to design a filter for the 10m band, you would first choose a cutoff frequency a bit

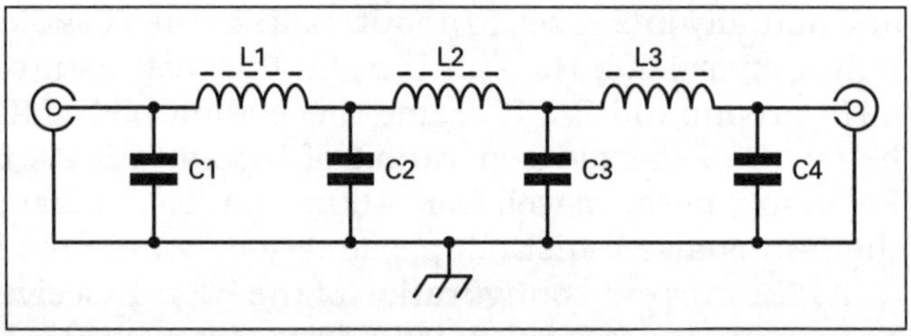

FIGURE 35: The 7th order PI low pass filter.

Filter type	C1	C2	C3	C4	L1	L2	L3
Butterworth	1417	5736	5736	1417	9.92	15.92	9.92
Chebyshev 0.1dB	3759.8	6673.9	6673.9	3759.8	11.32	12.52	11.32

Table 2: Normalised component values for Fc = 1MHz, Z = 50Ω

higher than 29.7MHz. About 31MHz would be ideal. Then, use the values from Table 2 to calculate the filter component values: C1 = 3759.8/31 = 121pF, C2 = 6673.9/31 = 215pF, C3=C2, C4=C1, L1 = 11.32/31 = 365nH, L2 = 12.52/31 = 404nH.

Filter Construction

The filters are built in the same manner as the filters in the BPF unit (page 21). Each filter is built on a strip of PCB laminate. I used 10cm x 2.5cm strips of double sided PCB. Miniature relays (Maplin N17AW or similar) are used for filter switching. It is not necessary to build a filter for all of the HF bands. A filter with a cutoff frequency of 22MHz can be used to cover the 15m and 17m bands. A 31MHz cutoff will cover the 10m and 12m bands. Six filters are enough to cover the nine HF bands; a seventh filter will be needed if you want to cover 6m as well.

A cutoff frequency of 2.1MHz was chosen for the 160m filter. Calculated values for the components are 1790pF, 3178pF, 5.39µH and 5.96µH. Standard value 1800pF capacitors can be used for C1/C4. I used a parallel combination of 1000pF and 2200pF to make 3200pF for C2 and C3. If an accurate capacitance meter is available, you can hand pick capacitors within their tolerance range to get even closer to the required values. The inductors for the 160m filter and all of the filters below 10MHz were wound on T50-2 toroid cores. The AL value of 49 can be used to calculate the number of turns required. Remember that these cores have a specified accuracy of ±5%. The exact value of inductance also depends on how evenly the turns are spaced on the core and, to a lesser extent, the thickness of the wire used. Maybe I tend to bunch the windings at one side of the core, but I usually find that my hand wound toroid inductors have slightly higher value than expected. 100~√(5.39/49) suggests 33 turns for L1 and L3. I found 32 turns was closer to the required value. 34.8 turns was calculated for L2; 34 turns was closest to 5.96µH on the inductance meter. I used 0.375mm enamelled wire for the 160m and 80m filter coils. Thicker wire can be used for the higher frequency bands where fewer turns will be required.

Measuring the filter performance with a signal generator and accurate power meter showed that the losses at 2MHz were unacceptably high, at around 2dB. Once again, cheap and nasty ceramic capacitors were found to be the cause. I had the same problem with the 160m filter in the BPF unit. It seems that physically small ceramic capacitors with high values around 1nF or more can be quite lossy at radio frequencies. After much swearing and digging in the junkbox, I eventually found some nice polystyrene capacitors with the correct values. Good quality mica capacitors should also be ideal for this circuit. If you are using ceramic capacitors, measure the filter losses carefully before you use the filter. Larger, ~10mm ceramic capacitors with a voltage rating of 100V or more should work very well in low pass filters. The rebuilt filter now shows

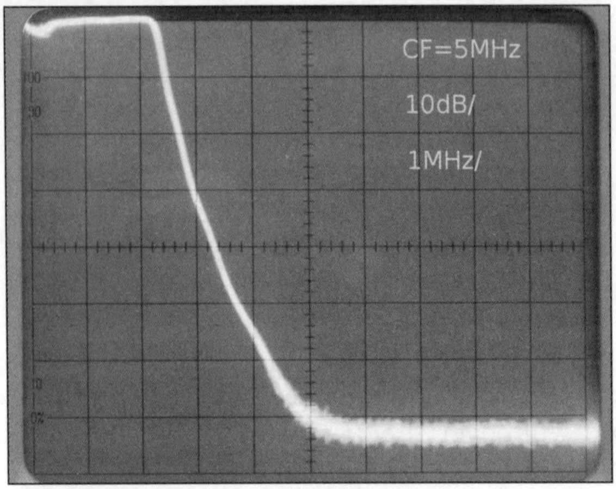

PHOTO 19: 160m low pass filter frequency response.

insertion loss of less than 0.5dB and I/O return loss of 18-20dB (SWR<1.3:1), as you would expect from a filter of this type. **Photo 19** shows the response of the 160m LPF for frequencies from 0 to 10MHz.

Now that we have solved the problems at the LF end of the spectrum, we will now tackle the 6m filter. Once again, we will use a 0.1dB Chebyshev. For a cutoff frequency of 55MHz, the required component values are 68pF, 120pF, 206nH and 227nH. More by accident than design, the capacitor values are the standard E12 values of 68 and 120. I used ordinary ceramic disc capacitors and T50-6 toroid cores for the inductors. These cores have an AL value of 40. $100x\sqrt{(0.206/40)}$ suggests that 7 turns is about right. No matter how carefully I distributed the turns around the toroid, the measured inductance was always more than the required value. 7 turns gave an inductance of 220-230nH; a bit too much for L1/L3 but exactly right for L2. I eventually used 6 turns for L1 and L3. This resulted in a measured inductance which was too low. But by adjusting the turns spacing so that the coil occupies about 66% of the core circumference, I was able to achieve the required value. See **Photo 20** for close-up details of L1, L2 and L3 for the 6m filter. This technique is very useful for fine tuning the value of low value toroid inductors. The insertion loss of the 6m LPF is less than 0.5dB. The filter response started to roll off at about 52MHz instead of the expected 55MHz. This is within the expected range of error. A couple of minutes spent tuning the values of L1 and L3 resulted in a perfect filter response. The assembled 6m filter is shown in **Photo 21**.

The only other filter that caused problems was the 10m filter. This showed a roll off frequency of just over 29MHz instead of the expected 31MHz. Testing the filter components showed that the 220pF capacitors used for C2 and C3 were actually just over 230pF. This is probably within the capacitor tolerance range, but it was just enough to de-tune the filter. Replacing these capacitors with new ones from another batch completely cured the problem.

I used 1mm wire for the 6m, 10m and 15/17m filters, 0.8mm wire for 20/30m and 40m ones. 0.375mm wire (Maplin YN86T) was used for the 80m and 160m filters. Table 3 shows component values for all of the LPF modules. **Figure 36** shows the switching

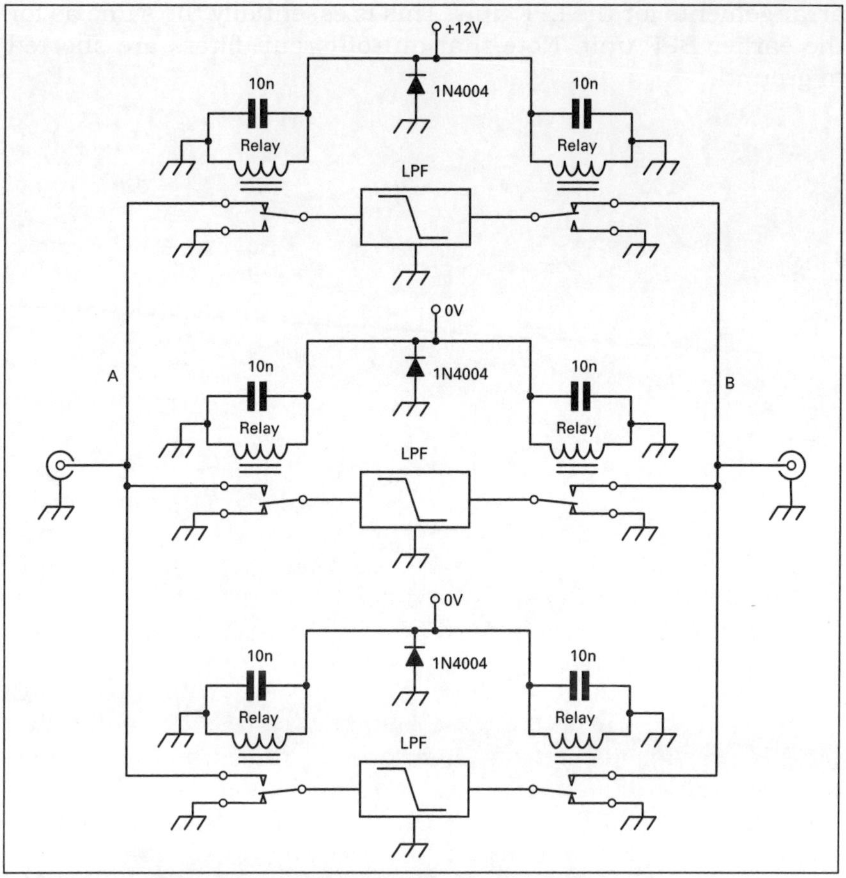

FIGURE 36: General arrangement of the low-pass filter switching.

PHOTO 20: The completed toroids for the 6m low pass filter.

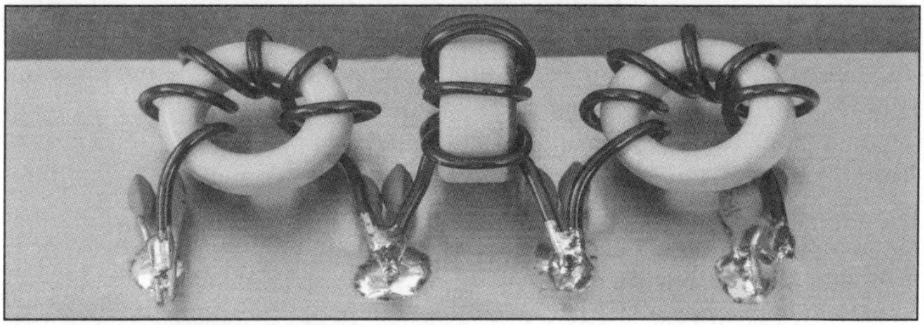

PHOTO 21: Completed 6m filter.

arrangements for the LPF unit. This is essentially the same as for the earlier BPF unit. Note that out-of-circuit filters are shorted to ground.

6

Starting the receiver section

BEFORE WE START building the receive front end of the transceiver **(Photo 22)**, perhaps we should examine how good the receiver front end needs to be? In late 2008, I did a survey of signals received by my HF doublet. This aerial is 40m long, which is approximately a half-wavelength on 80m. The feedpoint is 12m (40ft) above ground, the ends are a bit lower. The doublet is centre fed using 600Ω balanced feeder. This is a fairly typical amateur aerial. A bigger or higher aerial would probably add a few dB to received signal strength.

Over a period of one month, the strongest signals received during daylight hours were typically below -30dBm (7.1mV, or 1μW). Using the widely accepted convention of S9 = 50μV, 20log(7.1÷0.05) = 43dB above 50μV. So this -30dBm signal is fairly strong, at S9+43dB. At night, the stronger signals were usually below -20dBm (S9+53dB). The record for the strongest night-time signal was -13dBm from an East European broadcast station. This is equal to 50mV, or a rock crushing S9+60dB.

The strongest signals are often found in the 41m broadcast band, which is just above the European 40m amateur

PHOTO 22: Front end consisting of diode mixer, 10.7MHz diplexer and post-mixer amplifier.

band. **Figure 37** shows a snapshot of the RF spectrum from LF to 10MHz at 2030UTC on 21 March 2011. The strongest signal is at a level of -27dBm. Signals were a little stronger later in the evening, but even the biggest signals remained below -20dBm. It is worth noting that even the very strongest amateur signals are far below this level. The dominant signals are from short wave broadcast stations with transmitter power in the hundreds of kW and huge aerial arrays producing ERP measured in megawatts.

A perfect linear amplifier would produce no noise or distortion. The amplifier output signal would be an amplified but otherwise exact copy of the input signal. In the case of a single frequency sine wave input signal, the output signal would be a perfect sine wave and no harmonics would be produced. In the real world, even the best amplifiers are not perfectly linear. There is always some distortion that leads to the generation of harmonics and other spurious signals. Odd order (3rd, 5th, 7th...) inter-modulation distortion (IMD) products are particularly troublesome because they tend to produce in-band spurious signals at or near the frequency of wanted signals. The 40m band presents a worst-case scenario where there are many closely spaced and very strong broadcast stations close to the band edge. Third order IMD tends to increase as the cube of the input signal level. This means that for every 1dB increase of input power level, output 3rd order IMD will increase by 3dB. Increasing input power by 10 times (10dB) will increase 3rd order IMD by 1000 times (30dB)!

The strong signal handling capability of a receiver is defined by its input third order intercept point (IP3). This is a purely theoreti-

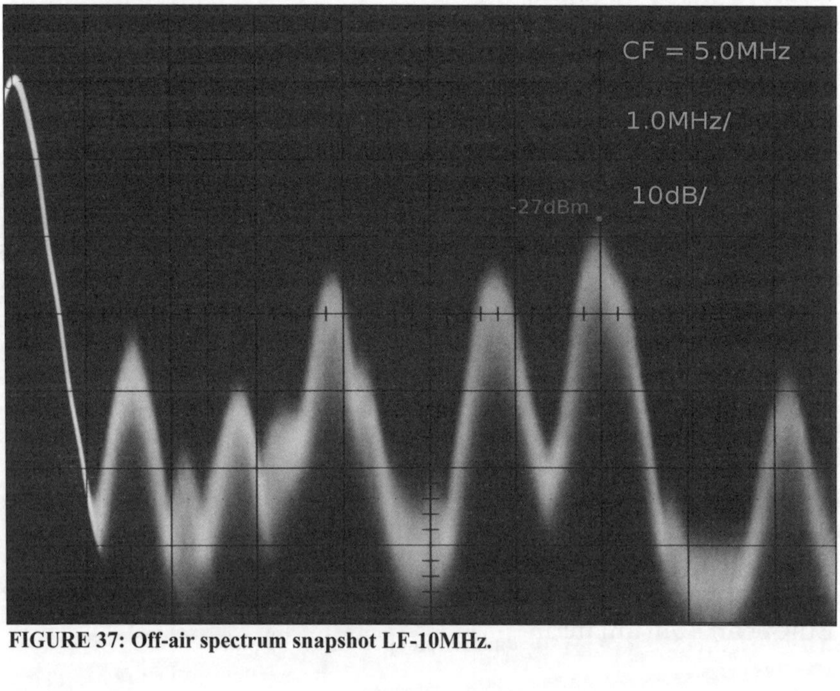

FIGURE 37: Off-air spectrum snapshot LF-10MHz.

cal power level where 3rd order IMD products have increased to a point where they are at the same level as normal input signals. We can calculate IP3 by using a two-tone test signal from a pair of signal generators. The two-tone signal of known level is applied to the input of the device under test (usually an amplifier or mixer) and the output signal is observed using a spectrum analyser or some form of frequency selective power meter. Once the relative levels of the two test signals and the 3rd order IMD products ((2*f1)–f2, (2*f2-f1)) are known, the IP3 can be calculated as IP3 = Pin + (IMDR/2) where Pin is in dBm and IMDR is the difference in level between the test tones and 3rd order IMD products. For example, if the device under test is an RF amplifier with gain = 10dB, each test tone is at a level of 0dBm (1mW) and 3rd order IMD is measured at 40dB below either test tone, the input IP3 is 0 + 40/2 = +20dBm. The output intercept point is 10dB higher at +30dBm. This test should be performed at an input level well below the 1dB compression point of the device under test.

Before we can start on IMD measurements, we will need a suitable pair of test oscillators. The two-tone test oscillators described in the January 2009 Homebrew have long since disappeared under the junk pile. **Figure 38** shows the schematic of one of the two test oscillators. The MPSH10 transistor is configured as a common-collector Colpitts oscillator. The output from the oscillator is taken via the 10MHz crystal. A 2SC495 CB driver transistor amplifies the 10MHz signal to a level of +20dBm (100mW). You can use a

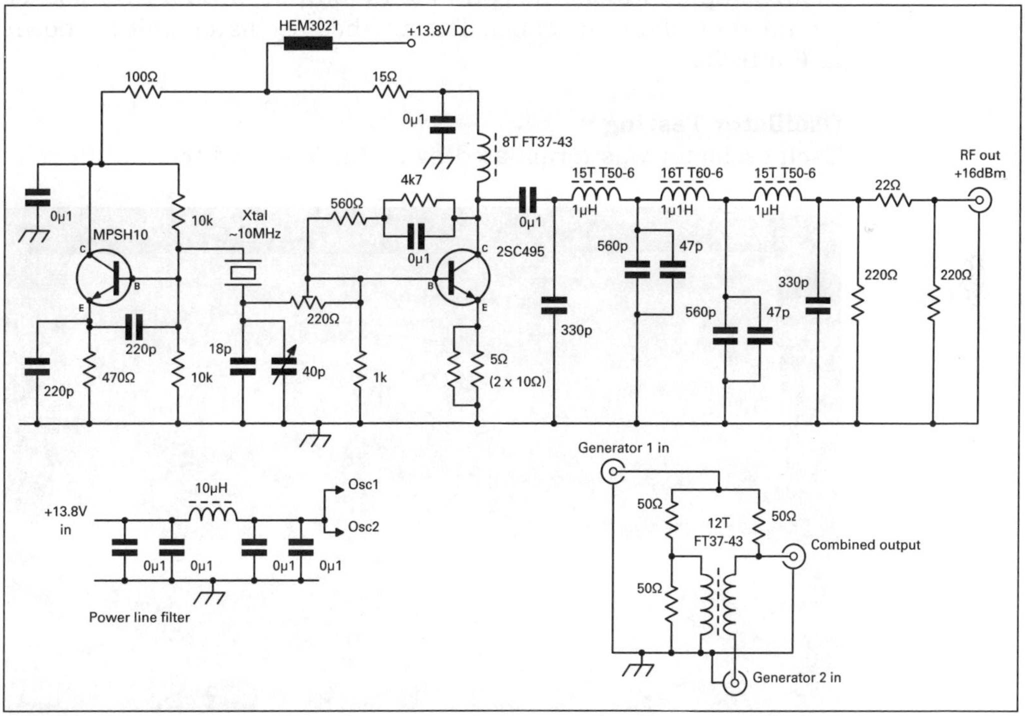

FIGURE 38: Colpitts test oscillator (top, two required), power line filter and output combiner.

VHF/UHF transistor like the 2N4427 or 2N5109 instead, but you will need to use a large clip-on heatsink. Power dissipation in this stage is around 1W. The output is via a 7th order LPF and a 4dB, 50Ω attenuator. The final output is +16dBm. The output level can be fine-tuned using the 220Ω pot attached to the crystal and an accurate power meter or oscilloscope.

Construction

The circuit is built dead-bug style on a strip of PCB laminate. The 2SC495 is mounted on the copper surface of the PCB using an insulated thermal mounting kit. A HEM3021 (Maplin N98AB) ferrite sleeve on the 13.8VDC supply wire to each oscillator module improves isolation between the two oscillators and reduces RF leakage via the DC supply wires. The output from the two oscillators is combined using a -6dB hybrid combiner. The -6dB hybrid allows the two oscillator signals to be combined while maintaining a high level of isolation between the two oscillators. This reduces IMD caused by one oscillator modulating the other. The combiner consists of three 50Ω resistors and a 1:1 balun transformer. I used a parallel pair of 100Ω, 1% tolerance metal film resistors for each 50Ω resistor. The 1:1 choke balun is 12 turns, bifilar wound (twisted pair) of 0.375mm enamelled copper wire on a FT37-43 ferrite toroid. The two oscillators are mounted in a home made aluminium box. Each oscillator is mounted vertically on a copper ground plane. DC power input is via a simple RF filter mounted close to the entry point for the DC supply wires. The two-tone RF output is via a BNC socket mounted on the front panel. The finished oscillator unit is shown in **Photo 23**.

Oscillator Testing

Each oscillator was terminated by a 50Ω load and the output volt-

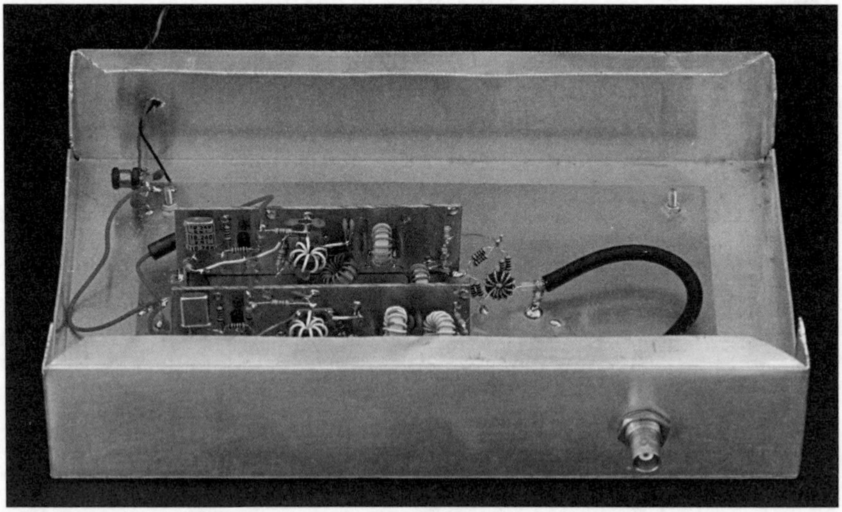

PHOTO 23: The completed two-tone test oscillator.

age was measured using a digital oscilloscope. The 220Ω pot was adjusted for an output level of exactly 2V peak or 4Vpp. $Vp^2/2R = 2^2/100 = 0.04W$ or +16dBm.

The LPF values are chosen for a cutoff frequency of 11MHz. I used 10.00MHz and 10.24MHz crystals because they were readily available and a 240kHz tone spacing is compatible with the 20kHz resolution bandwidth of my spectrum analyser. The 6dB power loss in the power combiner reduces the output of each tone to 10mW (+10dBm) or 40mW PEP for both tones. The peak instantaneous voltage value is twice the peak voltage from either generator. This happens when both generators are inphase. When both generators are out-of-phase the output voltage is zero. Bear this in mind when you are evaluating IMD levels. There is a 6dB difference between IMD values referenced to one tone and IMD values referenced to PEP. Low level tests of mixers and amplifiers usually specify IMD relative to either tone. High power testing of SSB transmitters and linear power amplifiers often specifies IMD relative to PEP.

Measuring the output from the combiner using a spectrum analyser confirms that the two tones are of equal amplitude and all IMD products are below the analyser noise floor at more than 70dB below either tone.

As a first practical test, the two-tone output was used to drive the input of a Class A power amplifier. The gain of this amplifier is 20dB. Measured output power is +30dBm (1W) for each tone, or 4W PEP. Rule-of-thumb calculations suggest that this amplifier will have an output 1dB compression point that is approximately the same as the DC input power, 13.8V x 3A = 40W (ish) or +46dBm. The output IP3 should be slightly more than 10dB above this value, +56dBm or better. Using a high power 40dB attenuator between the amplifier output and the analyser, the actual IMD products were measured at -58dB referenced to either tone. Input power is +10dBm for either tone, output power +30dBm each tone or 4W PEP. 10+(58÷2) = +39dBm input IP3 or +59dBm output IP3. Once again, the old rule-of-thumb proves quite accurate. These results confirm the very good reports on the signal quality from the new transmitter. At 8-10W PEP output, 3rd order IMD is still better than 50dB below PEP output. Distortion starts to rise rapidly at the 1dB compression point of approximately 20W.

Receiver

The first stage of the receiver is the RF BPF unit that has already been built and tested (page 21). Hopefully, I haven't put the cart before the horse by building this unit first and then testing its IMD performance several months later. My research into nonlinear behaviour of band pass filters led me to the excellent web page of Martein Bakker, PA3AKE **[7]**. This site is a treasure trove of detailed information about the H-mode mixer and general information about high dynamic range receiver design. Martein's measurements of 3rd and 4th order band pass filters using T50 type toroid cores shows that such filters will have an input IP3 of roughly +40dBm. This

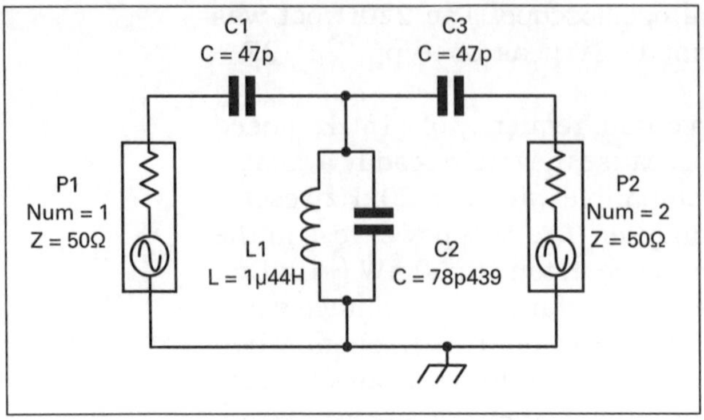

FIGURE 39: Single resonator BPF for test purposes.

easily meets my requirements. It seems that I won't have to put my new and rather expensive BPF unit in the dustbin after all.

For my own experiments, I built a very simple, single resonator BPF as shown in **Figure 39**. The -3dB bandwidth (BW) is about 1MHz and insertion loss is a small fraction of 1dB. The I/O coupling capacitors are 47pF disc ceramic types. Air dielectric trimmers would probably be better, but these particular ceramic capacitors seem to have high Q and good linearity. C2 is an Arco type compression trimmer. I have used these in 100W VHF amplifiers for some of our previous projects. The inductor is 19T of 0.375mm (not critical) wire on a T50-6 powdered iron toroid. Placing this filter between the two-tone oscillator and the spectrum analyser shows that there is little or no sign of any IMD at +10dBm (each tone). IMD at 60-65dB down suggests an IP3 of more than +40dBm. Using the Class A PA to boost the two-tone signal to +30dBm (each tone) shows 3rd order IMD at 40dB down. If we assume that non-linear behaviour of the toroid core follows the usual rules for IMD3, this suggests IP3 may be as high as +50dBm. After a long period operating at this level, the toroid core was slightly warm. The capacitors were still cold. Replacing the toroid with a 1.4µH inductor made from 1mm copper wire wound on 20mm plastic pipe reduced IMD products to 53dB below either tone. This suggests that most non-linear effects are due to the toroid core and not the capacitors. It is likely that a narrow bandwidth filter with a higher loaded Q will show higher levels of IMD than my simple circuit. However, I am satisfied that my bandpass filters will easily outperform any of the amplifiers or mixers that I am likely to use in the front end of my new receiver. State-of-the-art receivers using high level passive MOSFET mixers like the G3SBI H-mode mixer could potentially have an input IP3 of +40-50dBm. This would call for something more substantial than T50 cores in the BPF. Larger toroids like the T94 types used by PA3AKE or air core coils as described above would be suitable, but at a considerable cost in terms of size and weight.

Mixer

There are several obvious candidates for the first RF mixer. An old rig of mine uses a dual-bridge mixer as described in **[8]**. This mixer can achieve an IP3 of above +30dBm if sufficient LO drive is available. The standard diode DBM and the triple balanced mixer as used in the transmitter stage are also well proven high level mixers. I am also tempted to build a passive switching mixer using MOSFETs or high speed bus switch ICs. So many choices!

I evaluated several mixers before I settled on my final choice. The first front end will use a diode DBM with high local oscillator injection of up to +23dBm (200mW). The mixer will be followed by a diplexer with a Q of 5 and a push-pull transistor amplifier. The diplexer design was lifted from the January 2009 Homebrew project. The mixer and diplexer are shown in **Figure 40**. The mixer is based on a ring of four 1N5711 Schottky diodes, which were matched for forward voltage drop using the diode test function of my digital multimeter. The mixer transformers are each ten turns of trifilar wound 0.375mm enamelled copper wire. The three wires are twisted at 3 turns per cm. VE7BPO's website **[9]** has some good tips for transformer winding and diode matching.

The diplexer coils are 30 turns on a T50-6 toroid for the 3.88μH inductor and 5 turns on a T50-6 for the 150nH inductor. Because the diplexer component values are fixed, the component values are quite critical. The 150nH coil is bunched so that the five turns only occupy 50% of the toroid. If you don't have an accurate L/C meter, you might prefer to use variable inductors for the diplexer. As usual, the 50Ω resistors are a parallel combination of two 100Ω resistors.

I used the two-tone oscillator to test the diode mixer. The two-tone signal was applied to the RF port. The LO signal was provided by the DDS VFO (March 2010 Homebrew), a precision attenuator (April 2010 Homebrew) and a two stage amplifier (February 2011 Homebrew). This allows for LO drive level from microwatts up to +23dBm (200mW). With LO power in the 20-23dBm range and RF input of 0dBm (1mW) for each tone, third order IMD products were measured at 60dB below either tone. This indicates an input IP3 of +30dBm, which is an excellent result for such a simple and inexpensive mixer. Commercially made level-24 and level-27 mixer modules offer similar performance, but the price is 30-80 times the cost of my homebrew mixer.

I found that there was little or no difference in IMD output when the LO drive level was reduced from +23dBm to +20dBm. This suggests that there is little to be gained from increasing LO power beyond the +20dBm level. IMD increased slightly when LO drive was reduced to +17dBm and quite significantly as the LO power was

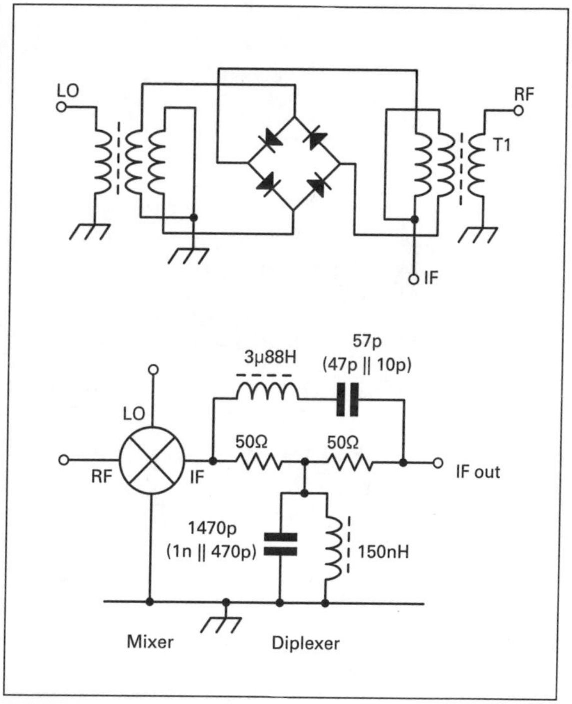

FIGURE 40: Mixer (above) and diplexer (below).

reduced to the 5-10dBm range. Interestingly, LO power has very little influence on conversion loss until the LO level is reduced to about 3dBm, which is only slightly more than the RF input level.

Testing

I used the IF/AF strip from a LF band receiver as a test bed for the receive mixers. The BPF unit was placed between the aerial and diode mixer RF input port. This combination has no RF or IF gain before the crystal IF filter. Performance was surprisingly good on the LF/MF bands. Band noise was easily detectable on 160/80/40 and several American stations were heard using this configuration. As you would expect, this system tends to be a bit deaf on the higher bands.

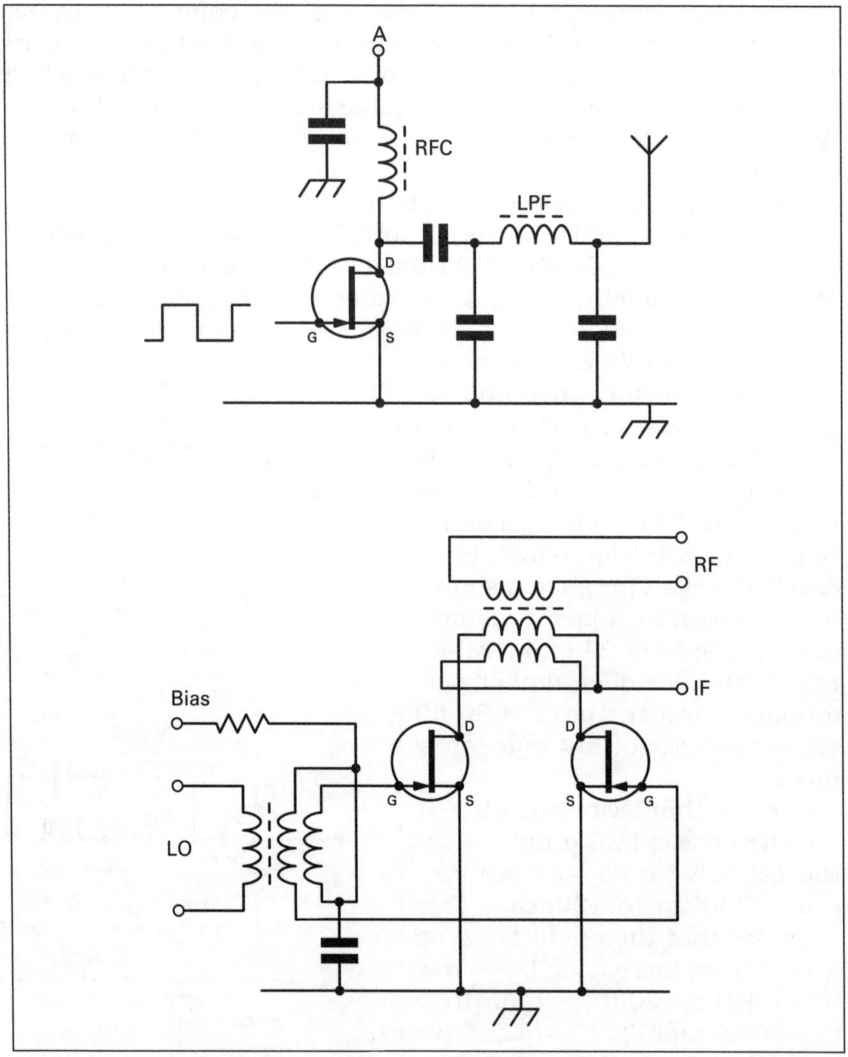

FIGURE 41: Direct conversion receiver mixer (top) and balanced version of the same (bottom) – see text.

Other Mixers

While I was testing the diode mixer, I also evaluated a simple MOSFET mixer design from a Russian magazine, top of **Figure 41**. I wanted to build a balanced version of this circuit so that the very strong LO signal would be isolated from the IF port. The result is shown at the bottom of Figure 41. I used two SMD MOSFETs stripped from a dead hard drive. The mixer performed very well on the LF bands, but not so well at HF, probably because of the limited switching speed and high gate capacitance of the MOSFETs.

The MOSFET mixer was tested using a 6Vpp sine wave at the MOSFET gates. Gate bias was initially set at 3VDC so that the LO drive swings from 0V to 6V. Tests with a dual trace oscilloscope confirm that the LO signal looks clean and symmetrical at the FET gates. I used a pair of 2N7000 switching MOSFETs in my test circuit. Both transformers are identical to the ones used in the diode mixer. Once again, a two-tone test signal of 0dBm each tone was used. IMD3 at the IF port was measured at 42dB below either tone. This suggests an IP3 of +21dBm, which is quite respectable, but not as good as I expected. There may be room for improvement if I use a square wave LO signal and perhaps some faster FETs or some high speed bus switches as used in recent versions of the H-mode mixer. I will return to this circuit in the near future. I would also like to try using four devices in both the ring and H-mode configurations. **Photo 24** shows my prototype MOSFET mixer.

PHOTO 24: Simple dual MOSFET mixer (see Figure 41).

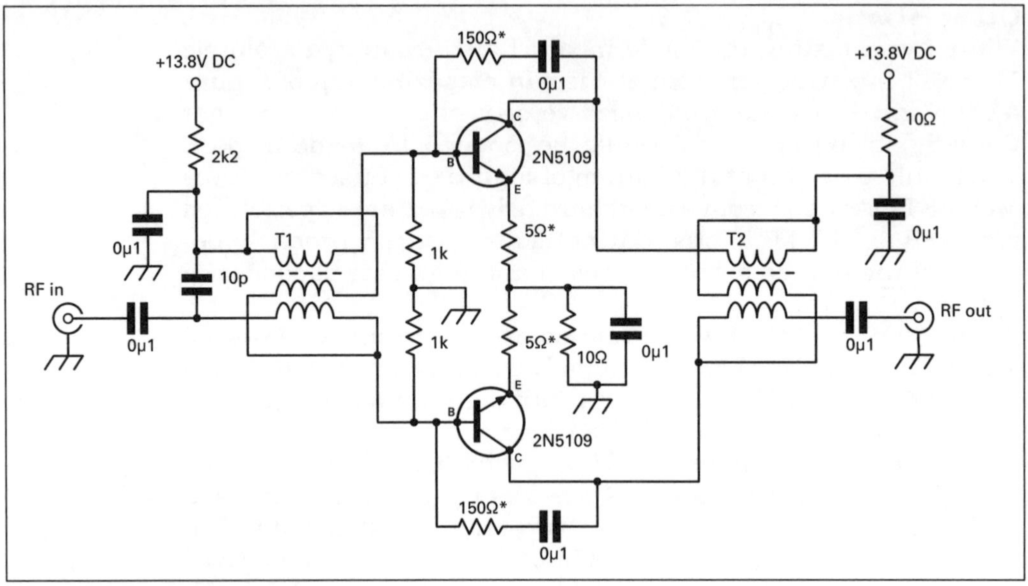

FIGURE 42: Post-mixer amplifier.

Post-Mixer Amplifier

The next stage in the chain is a post-mixer IF amplifier. As the mixer has a conversion loss, the mixer output IP3 is about 6dB below the input IP3. The diplexer loss is only about 1dB, so the post-mixer amplifier should have an input IP3 that is greater than +23dBm – and preferably significantly higher. I decided to use a push-pull amplifier, which is similar to the broadband amplifiers used in some of our recent projects. Each side of the amplifier is designed for a gain of 10dB and I/O impedance of 25Ω. The input and output balun transformers are simple 1:1 transmission line types. The amplifier schematic is shown in **Figure 42**. I used a pair of 2N5109 transistors biased for a high standing current of 50-60mA each. This is a DC input power of +32dBm, which should give a (rule of thumb) output IP3 of about 42dBm and input IP3 of about 32dBm, although I haven't measured the amplifier IMD yet.

Amplifier Construction

The amplifier was built on a strip of PCB laminate in the usual dead-bug style. Both transistors should be fitted with a large clip on heatsink. The smaller 10mm 44°C/W types are not big enough for this application. Finned 33°C/W or similar are recommended. T1 and T2 are both 10 turn trifilar wound (three twisted wires) on a FT37-43 toroid core. Pay close attention to the winding details in the schematic. The 5Ω resistors are made from two 10Ω resistors in parallel.

These 5Ω resistors and the 150Ω shunt feedback resistors set the amplifier gain at 10dB. If gain is a higher priority than output intercept point, you can increase the gain to about 15dB by using resistor values of 3.9Ω and 220Ω.

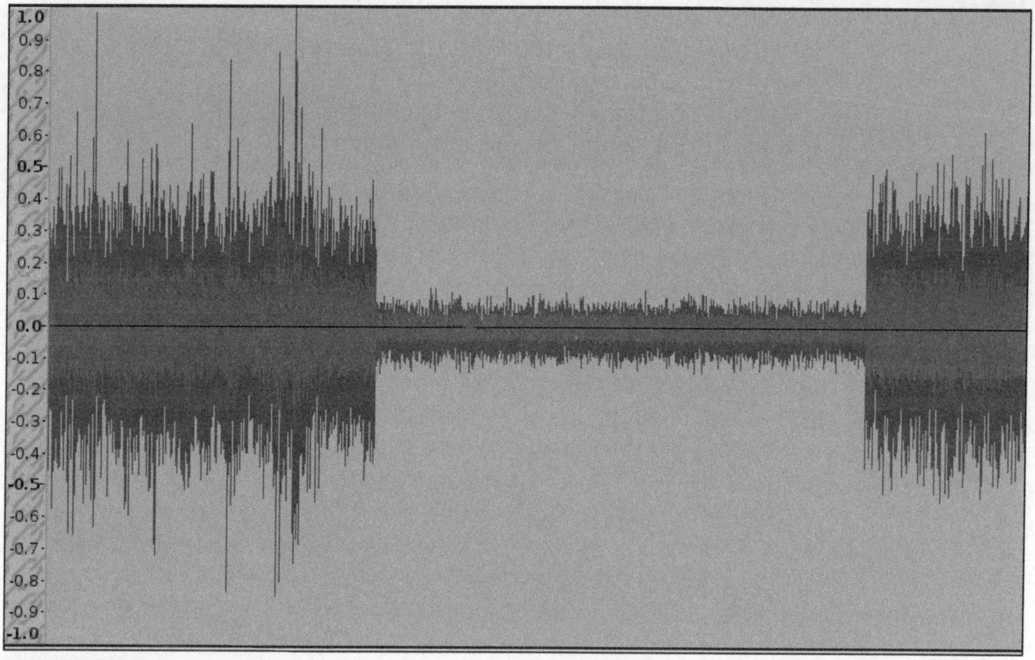

FIGURE 43: Receiver band noise test; the antenna was removed in the 'flat' portion (see text).

Amplifier Testing

Bench testing of the amplifier shows a gain of exactly 10dB, falling off to 9dB at 32MHz and 7dB at 60MHz. Input return loss is better than 16dB (SWR 1.4:1) from LF to 30MHz and better than 14dB (SWR 1.5:1) up to the low VHF region. This is ideal for the intended application as a post-mixer IF amplifier. I would have liked to see slightly better performance from 28MHz to 52MHz so that the same design could also be used as the RF amplifier.

The receiver test rig was rearranged in the following order: BPF unit – mixer – diplexer – post-mixer amplifier – IF strip. Receiver sensitivity is now much improved on the HF bands. The very first signal heard was veteran homebrew expert Harold, W4ZCB on 15m. Recordings of this test can be found at **[10]** and **[11]**.

No noise figure measurements were performed, but the receiver sensitivity is good enough to hear the band noise on all bands up to 10m. **Figure 43** shows the received noise on a quiet frequency on 15m. This is a six second recording. The receive aerial was disconnected for three seconds. Average band noise is more than 10dB above the receiver noise floor, peak noise spikes are about 20dB above the noise floor. Not bad for a rig with no RF amplifier.

[7] www.xs4all.nl/~martein/pa3ake/hmode/
[8] *Solid State Design for the Radio Amateur*, Hayward, DeMaw. ARRL.
[9] www.qrp.pops.net/xmfr.asp
[10] http://homepage.eircom.net/~ei9gq/ W4ZCB-15M.mp3
[11] http://homepage.eircom.net/~ei9gq/N2IR.mp3

7

IF amplifiers, a product detector and an audio amp

HISTORY - THE 'SUPERHET (supersonic-heterodyne or superheterodyne) receiver was developed during the First World War. Edwin Armstrong is generally credited with the invention of the superhet. Armstrong also invented the regenerative receiver, super-regenerative receiver and developed the first FM radio systems. Before the invention of the superhet, radio receivers used the newly invented triode valve as RF amplifiers and audio amplifiers. The amplification provided by the triode valve resulted in a huge improvement in the sensitivity of radio receivers. Valves from this era had useful gain at audio frequencies and some gain at LF/MF, but they had little or no gain at shorter wavelengths. Reception of weak signals was only possible at the low end of the RF spectrum using TRF (tuned radio frequency) receivers. These receivers had multiple stages of RF amplification. Each stage was tuned to the frequency of the incoming signal so that the selectivity of the receiver was defined by the bandwidth of the RF amplifier stages. Receivers from this era suffered from poor sensitivity and a lack of selectivity. The superhet provided a perfect solution to both problems. Short wave signals could be converted to a lower (intermediate) frequency where they were easily amplified by the triode valves of the day. The IF amplifiers also provided the main selectivity of the receiver. As the intermediate frequency was fixed at a constant value, there was no need to retune the IF when the receiver was tuned to a new frequency. The following text from Armstrong's patent application gives a very clear description of the superhet receiver.

"This new method of reception consists in converting the frequency of the incoming oscillations down to some predetermined and lower value of readily amplifiable high frequency current and passing the converted current into an amplifier which is adjusted to operate well at this predetermined frequency. After passing through the amplifier, these oscillations are detected and indicated in the usual manner."

The full text of Armstrong's US patent is available on the web **[12]**. This was one of the most important developments in the history of radio. The superhet allowed HF and VHF signals to benefit from the high gain and good selectivity of IF amplifiers operating at relatively low frequencies. The superhet had been the dominant radio receiver architecture for almost a century. The great majority of TV, broadcast radio, commercial and amateur radio receivers in use today are superhets.

The back end of a superhet is essentially the same as a TRF set. A set of tuned amplifiers is followed by the detector and audio amplifier stages. The superhet front end adds a frequency changer consisting of a local oscillator and mixer.

A superhet front end usually has one or more RF tuned circuits that are used to suppress the IF image and other unwanted signals like direct IF breakthrough. The superhet may also have one or more stages of RF amplification. Some of the more complicated superhet receivers have several stages of frequency conversion from RF to first IF, first IF to second IF and so on. Armstrong's superhet patent proposes a double conversion receiver with a first IF of 1MHz and a second IF of 100kHz.

Early triode valves had relatively large inter-electrode capacitance. The anode to grid capacitance was particularly troublesome because it results in negative feedback that limits gain at higher frequencies. This capacitance is effectively multiplied by the gain of the valve (the Miller effect). Later developments in valve design improved anode to grid isolation by placing a fourth electrode (screen grid) between the anode and control grid. Many textbook references describe Miller effect in triode valves, but any inverting amplifier with capacitive coupling from output to input will show increased capacitance (Miller capacitance) due to amplifier gain. Semiconductor devices designed for use in VHF/UHF amplifiers have low values of collector to base capacitance (in the case of bipolar transistors) or low gate to channel capacitance (FETs).

Some circuit configurations are not prone to increased input capacitance due to Miller effect. One such configuration is the common base amplifier. This is a non-inverting amplifier with relatively low gain and good isolation between the input and output circuits. The common base configuration was widely used in early transistor based VHF/UHF TV and radio receivers. The FET equivalent is the common gate amplifier. The valve equivalent is the grounded grid (GG) amplifier.

The 'cascode' amplifier is one of those magic circuits that provides all the advantages of a common emitter/source/cathode amplifier and none of the disadvantages. The cascade is a two stage amplifier consisting of a common cathode (or, depending on the device, emitter or source) amplifier as the first stage and a common grid/base/gate amplifier as the second stage. This configuration offers a relatively high input impedance, excellent input/output isolation (low reverse gain), high gain and/or wide bandwidth. **Figure 44** shows a cascade amplifier based on a pair of triode valves. The simplified circuit at A shows a cascade amplifier with DC bias and coupling components excluded. The first stage is a common cathode amplifier with the input signal applied to the grid and the output signal taken from the anode. The second stage is a grounded grid amplifier with the signal from the first stage applied to the cathode and the output taken from the anode. You would expect a two stage (cascade) amplifier to have a very substantial power gain somewhere in the region of 2x20dB = 40dB or perhaps even more. But because of the severe mismatch between the high output impedance of the first stage and the low input impedance of the second stage, the gain of a typical cascode amplifier is reduced to about 20-25dB. It is this reduced gain and improved I/O isolation due to the buffering action of the second stage that leads to greatly reduced Miller effect and, consequently, lower input capacitance. The circuit at Figure 44B shows a complete cascode amplifier with a tuned output stage, as typically used in an IF amplifier. This circuit was often based on a double triode valve like the ECC88 that had two triodes in a single envelope. Gain control is applied by varying the voltage applied to the grid of the second stage. As we will see later, this circuit is not just of historic interest. Many modern electronic circuits are based on cascode amplifiers using transistors and FETs in both discrete and

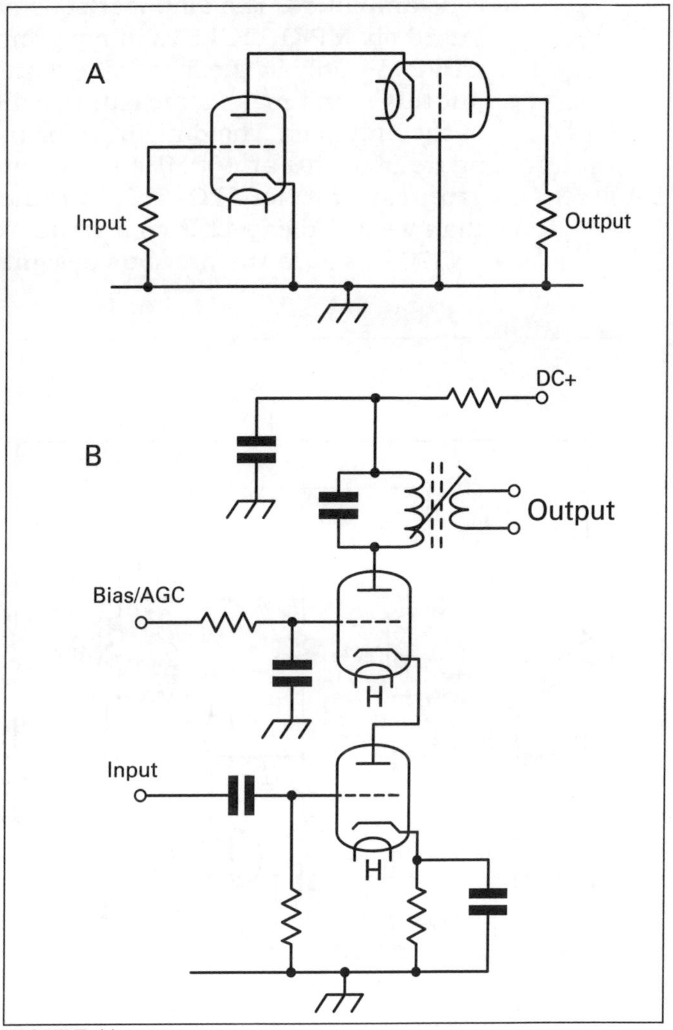

FIGURE 44:
(A) Basic 'cascode' configuration, omitting supply and bias components.
(B) Practical cascode amplifier with tuned output stage, as typically used in an IF amplifier.

integrated circuit form. The FET equivalent of the triode cascode is identical to Figure 44B except that JFETs are used instead of the triode valves. A cascade made from a pair of bipolar transistors will have slightly different biasing arrangements. It is also possible to make a cascode amplifier using a mix of different devices. One popular arrangement uses a JFET in the first (bottom) stage and a bipolar transistor in the second (top) stage. Dual-gate (DG), four terminal devices like the DG MOSFET and DG GaAsFET are internally configured as cascode structures having a single channel and two gates.

A JFET Cascode IF Amplifier

This project is an IF amplifier strip consisting of three JFET cascode amplifiers. The amplifier schematic is shown in **Figure 45**. I used six MPF102 JFETs in my prototype circuit. I did try using a BF245A only in the first stage (bottom half of the first cascode) but this didn't make a measurable difference to the performance of the amplifier. The datasheet for the BF245A specifies a noise figure of 1.5dB at 100MHz for a common source amplifier with gate resistance of 1kΩ. This is rather good performance, better than we could expect from a standard bipolar transistor like the 2N3904 used in the previous design.

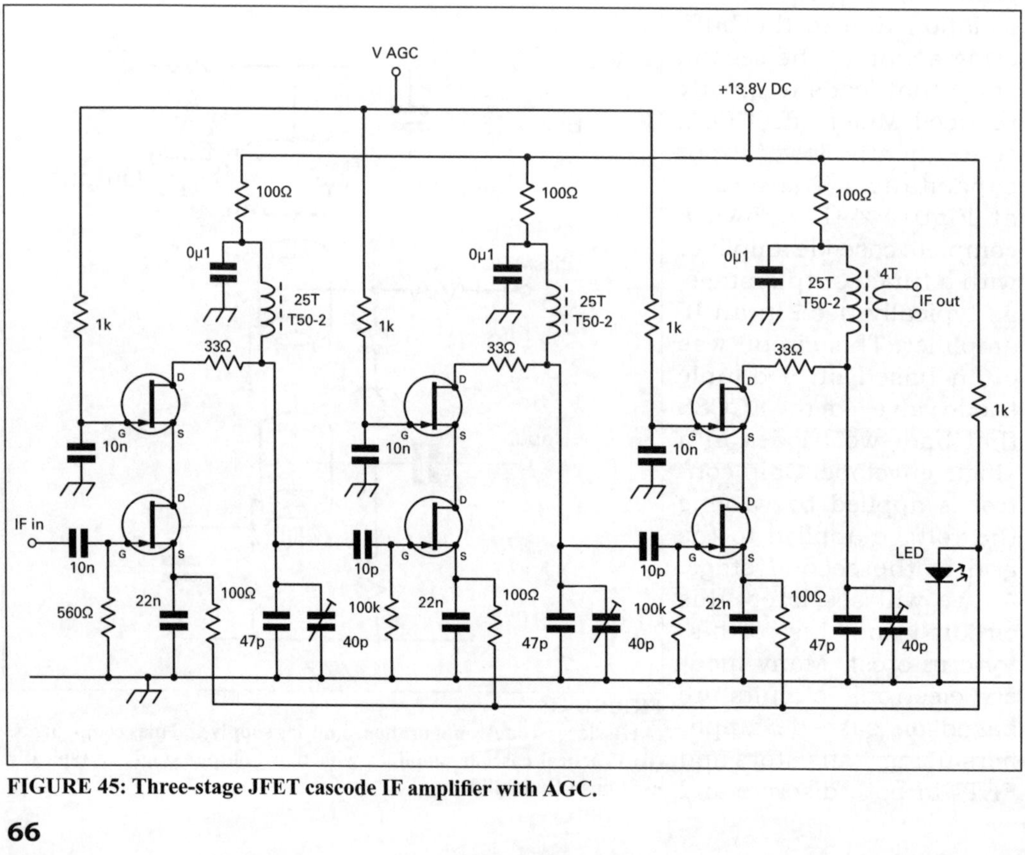

FIGURE 45: Three-stage JFET cascode IF amplifier with AGC.

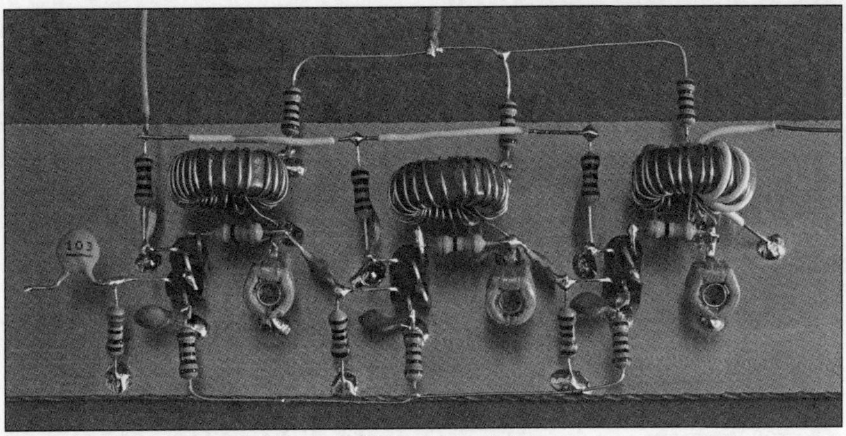

PHOTO 25: JFET cascode IF strip built dead-bug style.

Cascode IF Amp Construction

The amplifier was built in the usual dead-bug style on a strip of copper PCB laminate **(Photo 25)**. Each cascode pair of FETs was soldered together as shown in **Photo 26**. I started by building a single cascode stage to do the initial tests and measurements. Gain at 10.7MHz was measured at 20-24dB for each cascode pair. +5 to +6V on the AGC input gave maximum gain. AGC control range is very limited unless some positive bias is placed on the source of the lower FET. I used an LED as a poor man's Zener diode. The forward voltage drop across the LED is about 2.3V. The inductor for the L/C tuned circuit in each amplifier is 25 turns of 0.375mm (not critical) enamelled copper on a T50-2 powdered iron toroid. The inductor in the final stage also has a 4 turn coupling winding for driving the 50Ω input of the product detector.

The complete IF strip shows a gain of 70dB at 10.7MHz, falling to around 0dB as the AGC line voltage approaches zero volts. **Table 4** shows amplifier gain vs control voltage.

AGC V	Gain
6.00V	70dB
1.93V	60dB
1.33V	50dB
1.07V	40dB
0.76V	30dB
0.52V	20dB
0.29V	10dB
0.07V	0dB

TABLE 4: IF strip gain vs AGC voltage.

PHOTO 26: Two MPF102 JFETs soldered in cascode form.

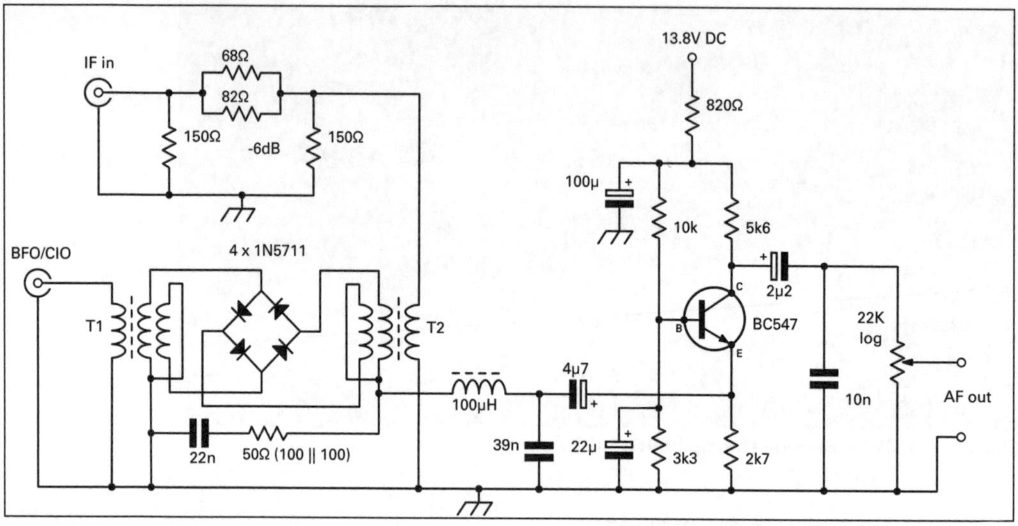

FIGURE 46: Product detector and audio buffer unit. See text for details of T1 and T2.

A simple diode double balanced mixer (DBM) was used as the product detector (PD). This is followed by a common-base audio amplifier. The schematic of the PD and audio unit is shown in **Figure 46**. T1 and T2 in the product detector are each 10 trifilar turns of 0.375mm enamelled copper on a FT37-43 ferrite toroid. I used four 1N5711 Schottky diodes that I matched for forward voltage drop using a digital multimeter.

I tested the IF strip using the RF front end (chapter 6), the BPF unit plus the audio power amplifier shown in **Figure 47**. I used a 10k pot across the 13.8V power supply to provide the AGC line voltage for these tests. The amplifier performed very well. Many DX signals were heard on the HF bands without the benefit of an RF amplifier. The IF gain seemed slightly excessive for most situations, so I placed a 6dB attenuator between the IF strip output and the input of the PD. The attenuator is shown in the PD schematic, Figure 46.

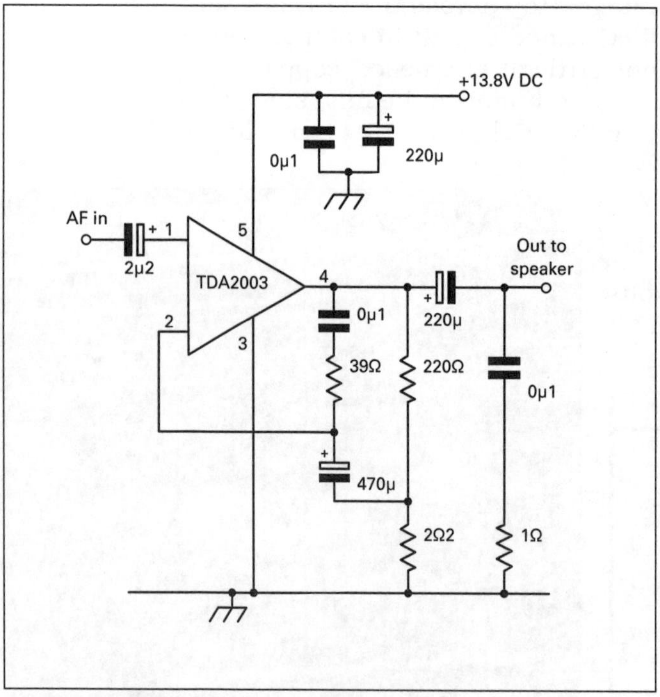

FIGURE 47: Simple yet effective audio amplifier based on the inexpensive TDA2003.

IC IF Strip

The next project is another IF strip based on the Analog Devices AD603 variable gain amplifier IC. Most IF amplifiers use some non-linear element for gain control. The AGC system of most FET and valve IF amplifiers varies the gain of each controlled stage by varying the bias voltage. This method often leads to reduced linearity of the IF amplifier when input signals are at a high level. The AD603 takes a different approach, using a fixed gain RF/IF amplifier preceded by a variable attenuator. The broadband RF amplifier uses an opamp style negative feedback network. This offers several possible configurations for the fixed gain stage. The default configuration gives a maximum gain of around 40dB. If the output and feedback (FB) pins of the IC are strapped together, the gain is reduced to 31dB. The gain control system gives a control range from 0dB to 42dB. The really nice feature of this amplifier is that the relationship between control voltage and gain/attenuation in dB is linear. The gain of the amplifier is controlled at a rate of 40dB per volt or 25mV per decibel of gain reduction. This results in an amplifier with accurate and predictable gain. It also provides smooth and accurate AGC action. Because the passive attenuator precedes the amplifier, there will never be a situation where very strong signals are applied to the amplifier input. Strong signals within the IF passband will cause the AGC to increase the attenuation value. This is an ideal situation because the amplifier shows maximum linearity at minimum gain when input signals are at a high level. **Figure 48** shows my test circuit for a single stage AD603 amplifier. My first prototype didn't

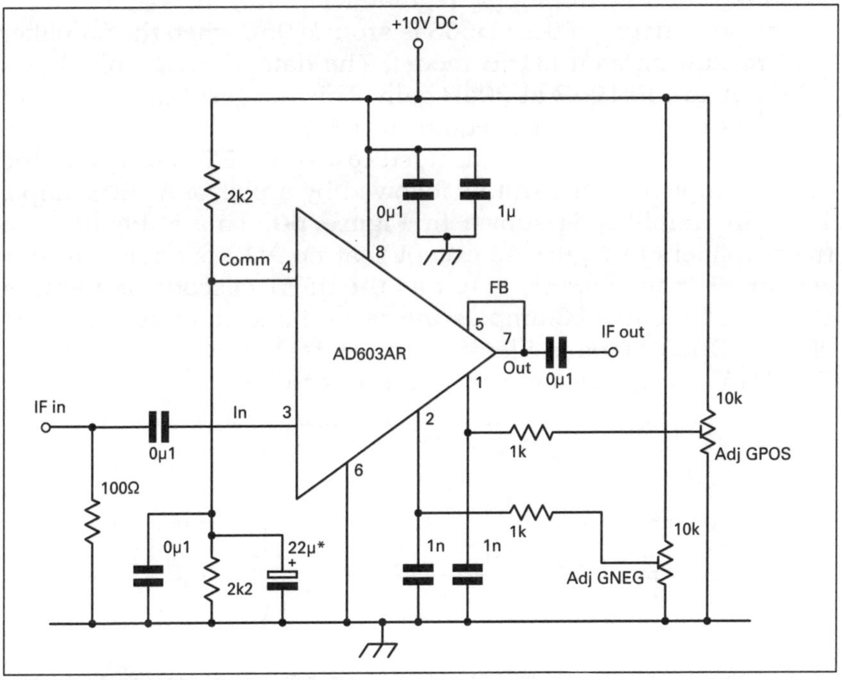

FIGURE 48: Test circuit for a single stage AD603 amplifier. Note tantalum capacitor (*).

PHOTO 27: Initial test setup for the AD603 IF amplifier (see Figure 48).

have a link between pin 5 and 7 so that the gain was 40dB and the control range was 40dB to 0dB. With the FB link in place, the gain range is +31dB to -11dB. I used a surface mount AD603AR in a dead bug circuit. This amplifier is shown in **Photo 27**. The measured performance of the amplifier was exactly in line with the data sheet figures. The amplifier has a gain of 40dB (without FB link) and control range of just over 40dB. The 100Ω resistor across the input gives an RF/IF input impedance close to 50Ω. The amplifier normally uses a split power supply of ±5V. As this would be quite inconvenient, I biased the common input (pin 4) to half the supply voltage so that a single-ended 10V DC supply can be used. **Figure 49** shows the 10V DC supply regulator.

The noise figure of the AD603 is around 9dB when the amplifier is at maximum gain (31dB mode). The data sheet graph shows NF rising to NF=19dB at 20dB gain. This suggests that input NF degrades at a rate that is equal to the gain reduction value. To overcome this limitation, I decided to use a FET cascode as the first IF amplifier. This will be followed by a pair of AD603 amps. The FET amplifier is shown in **Figure 50**. This is identical to the amplifiers in Figure 45 except that no AGC is applied to this version. The finished IF strip has the JFET cascade as the first stage, and the AD603 amp as the second and third stages. Each of the AD603 amps is identical to the test circuit in Figure 48. The AD603 amplifiers are shown in **Photo 28**.

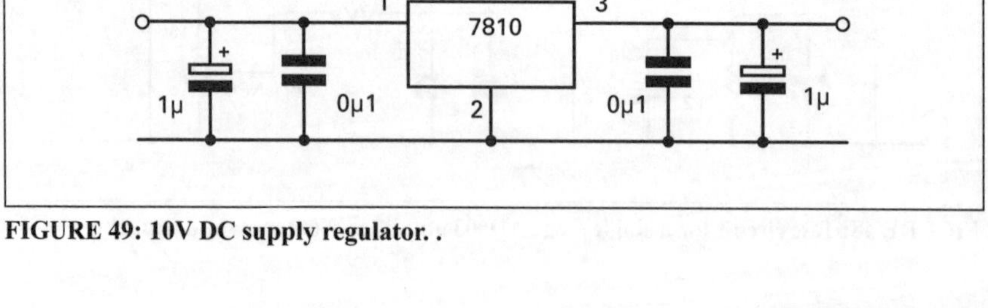

FIGURE 49: 10V DC supply regulator. .

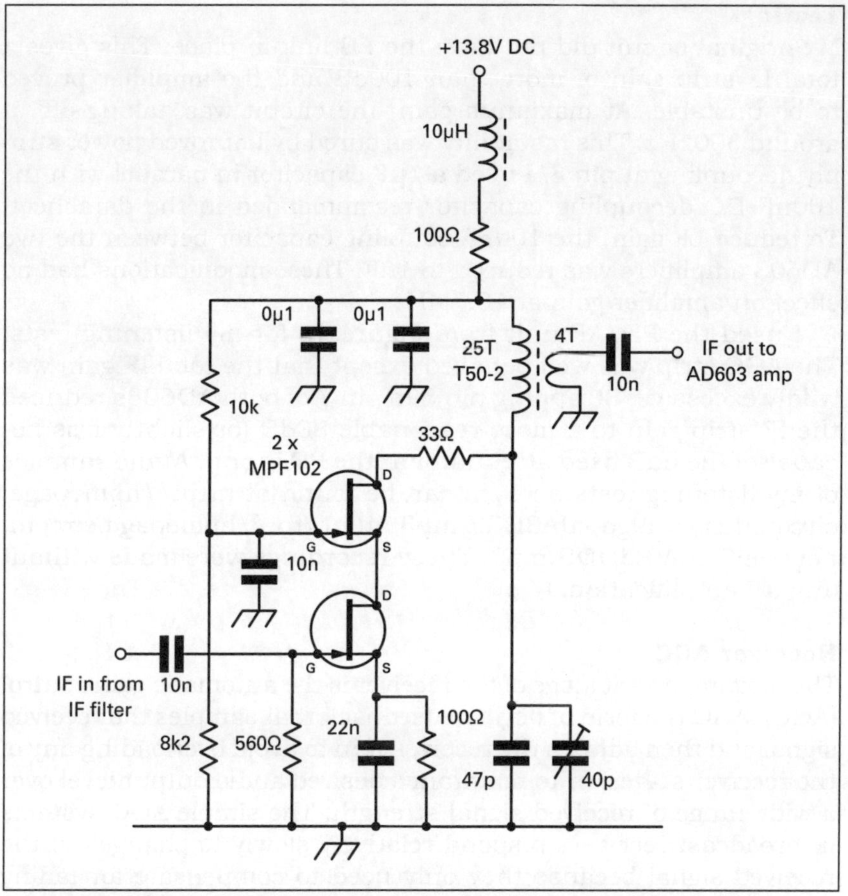

FIGURE 50: FET IF amplifier front end that feeds the AD603 strip of Photo 28 (see text).

PHOTO 28: Two-stage AD603 IF amplifier..

Testing

My original circuit did not have the FB link in place. This gives a total IF strip gain of more than 100dB and the amplifier proved to be unstable. At maximum gain, the circuit was 'taking off' at around 300kHz. This instability was cured by improved power supply decoupling at pin 8. I used a 1µF capacitor in parallel with the 100nF DC decoupling capacitor recommended in the datasheet. To reduce LF gain, the 100nF coupling capacitor between the two AD603 amplifiers was reduced to 1nF. These modifications had no effect on amplifier gain at 10.7MHz.

I used the PD/AF unit from Figure 46 for my listening tests. The AGC strip was well behaved except that the total IF gain was wildly excessive. Strapping pin 5 to pin 7 of both AD603s reduced the IF strip gain to a more reasonable 80dB (or slightly less because of the 6dB fixed attenuator at the PD input). Audio samples of my listening tests on 17m can be found at http://homepage.eircom.net/~ei9gq/ JA3EGZ.mp3 and http://homepage.eircom.net/ ~ei9gq/VE3DDB.mp3. These recordings were made without any RF amplification.

Receiver AGC

The next important stage of the receiver is the automatic gain control (AGC). AGC is a form of negative feedback that samples the received signal and then adjusts the receiver gain to avoid overloading any of the receiver stages, or to maintain a desired audio output level over a wide range of received signal strength. The simple AGC systems in broadcast receivers respond relatively slowly to changes in the received signal because they only need to compensate for fading (QSB). They aren't expected to cope with sudden, dramatic changes in received signal strength as might be encountered when working both weak and strong stations at the same time. The rapidly changing nature of SSB and CW signals is particularly challenging for the AGC system designer.

There are many different methods of applying AGC to a receiver. Audio-derived AGC samples the output from an AF amplifier in the receiver and adjusts the receiver gain to achieve the desired level. IF derived systems sample the output at one of the final IF amplifiers to determine received signal strength. In a digital (SDR) system, the AGC may be implemented entirely in software, although in some SDR systems, RF gain control is applied to the receiver front end to prevent overloading of the ADC at a later stage in the receiver.

Some direct conversion receivers use AGC that is entirely audio-derived. As there is no IF amplifier, it is not possible to apply gain reduction at an IF stage. As many direct conversion receivers don't have any RF amplifier either, the only place where AGC is easily applied is in the high gain audio amplifier chain. This is notoriously difficult to do without causing noticable noise and distortion.

Superhet SSB and CW receivers can sample the signal from

an audio stage after the product detector and then use the level of this signal to apply AGC to the IF and/or RF amplifier. This arrangement can be made to work very well **[13]**. This system is much less prone to the various pops, clicks and distortion which often afflict the simple DC receiver with audio AGC.

IF-derived AGC systems usually offer better dynamic performance than AF-derived systems. Audio-derived systems have a limited 'attack' time, especially if the received signal produces a relatively low frequency audio signal at the product detector, as is often encountered with SSB or low-pitch CW signals. The AGC attack and decay times are critically important for SSB and CW reception. The attack time is the period between the arrival of a strong signal and the application of AGC. The receiver will operate at close to maximum gain while receiving a weak DX signal. If a very strong signal appears on the same frequency, it may be necessary to reduce the receiver by a very large amount in a short time period - perhaps as much as 40-50dB of gain reduction will be required, typically in less than a couple of ms. To prevent the gain control system from tracking the modulation envelope of the received signal, most AGC systems have a relatively slow decay time. The ratio of attack time to decay time is typically about 1000:1.

The gain reduction during the attack period must happen without causing any noticeable distortion of the received signal. Obviously the band noise will be greatly reduced by the heavy application of AGC. This tends to be masked by the strong received signal that caused the AGC action. There are a number of problems that affect the operation of analogue AGC systems. Slow attack will cause a loud surge or click on the leading edge of strong signals. This is particularly apparent on strong CW signals. Another common problem is AGC overshoot, where excessive gain reduction is applied. **Figure 51** shows a sample of a received signal with slight delay on the attack and some signs of overshoot. There is no perfect solution to these problems. Because of the time delay

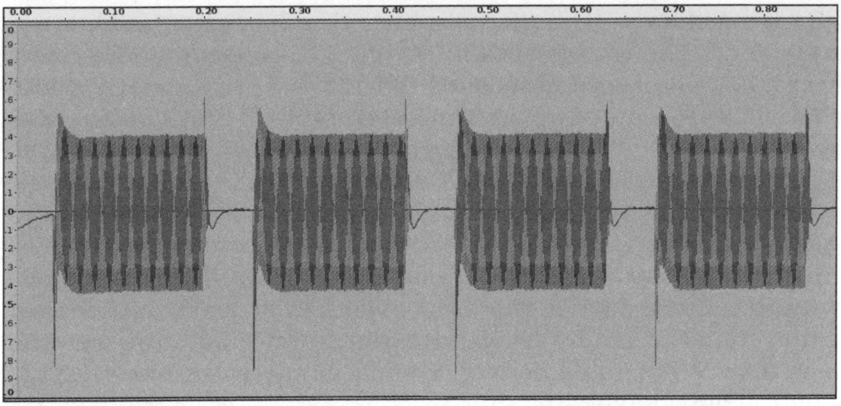

FIGURE 51: Receive signals with slight delay on the attack and slight overshoot.

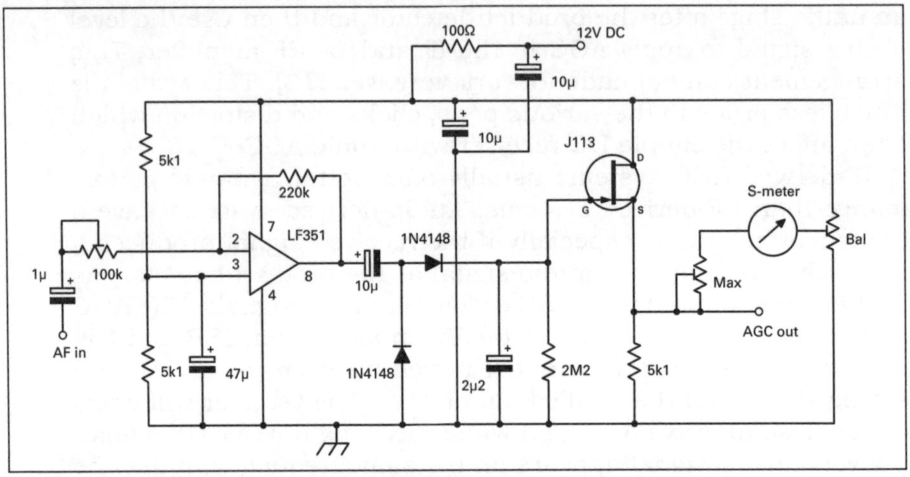

FIGURE 52: A simple audio-derived AGC system.

while the signal is propagating through the narrow IF or audio filter, it is not possible for the AGC to react to a strong signal until some time after it arrives at the receiver input. By this time, some of the later stages of the receiver may already have been driven into limiting. One partial solution is to use an entirely separate IF or AF amplifier for the AGC system – but this must have a greater bandwidth than the main IF or AF filters, so it will tend to respond to off-channel signals as well as the wanted signal. Most analogue receivers use just a single IF chain with carefully chosen time constants for the AGC attack and decay. **Figure 52** shows a simple audio-derived AGC system that I have used in several receiver projects. Receive audio is taken from the last audio stage before the AF gain control. The hot end of the AF gain pot is usually a convenient point from which to extract the audio signal. This signal is amplified by an opamp and then rectified to provide the DC signal used to generate the AGC control voltage. Because the opamp has a relatively low output impedance, the 2.2μF capacitor is charged very quickly when a strong signal is received. This provides the fast AGC attack time of a few ms. The FET is used as a DC buffer amplifier that has a very high input resistance. The RC network in the gate circuit of the FET has a time constant of several seconds. This provides a slow AGC decay time. The DC output at the source of the FET provides the AGC voltage and also drives the S-meter.

All of the superhet SSB/CW receivers that I have built in the past have used audio-derived AGC. I built several audio derived AGC systems for the LF/MF receiver. Most of them used FET-input opamps as precision rectifiers and DC buffers. All of them worked to some degree, one version worked very well, but not any better than the simple system described previously. You can find a sample of the audio derived system in operation here: http://tinyurl.com/y85kok4.

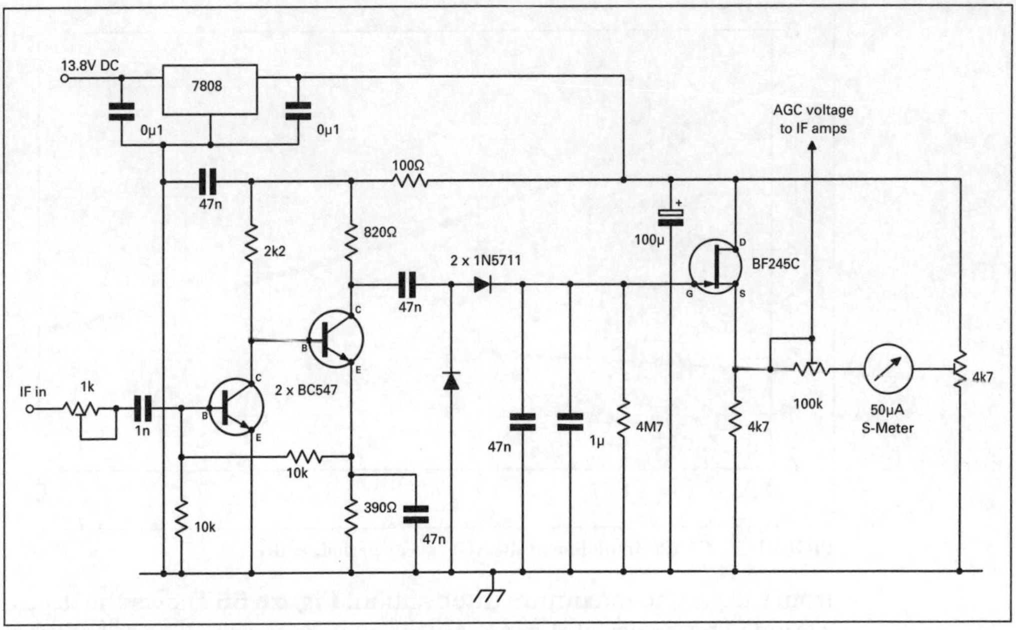

FIGURE 53: Practical IF-derived AGC system.

Just for a change, I decided to have a go at building an IF-derived AGC system. The new IF derived system consists of a two-stage 10.7MHz IF amplifier, which is followed by a voltage doubling rectifier and a FET DC buffer. The 10.7MHz amplifier input is connected to the 50Ω output of the receiver IF strip at the input to the product detector. The new 10.7MHz amplifier is a direct-coupled, two stage, common emitter type and in the presence of a strong received signal it produces several volts p-p of IF signal at the output. This amplifier is followed by a voltage doubling rectifier. The 1N5711 Schottky diodes have a lower forward voltage drop than the IN4148 diodes used in the earlier circuits. **Figure 53** shows the schematic of the AGC system. The RC network in the gate of the FET buffer has a time constant of just under five seconds. **Figure 54** shows a QUCS simulation of the attack period. The horizontal scale is 1ms/div. It takes less than 1ms to go

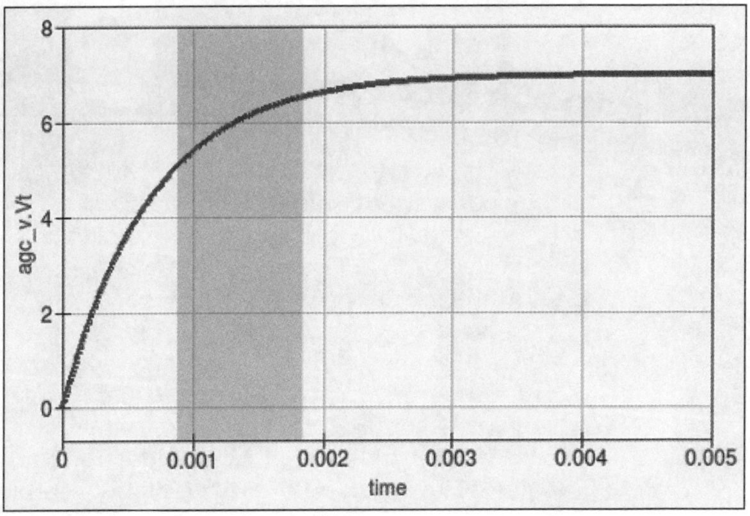

FIGURE 54: QUCS simulation of the AGC attack period, 1ms/div.

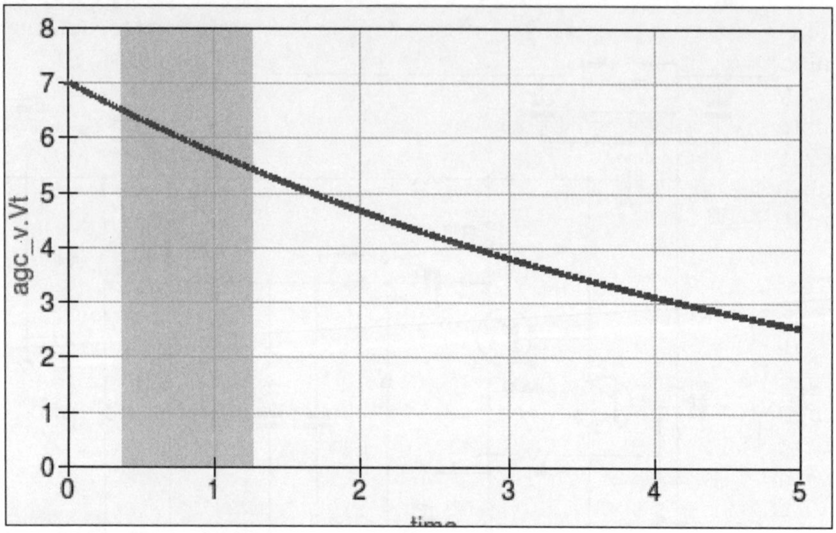

FIGURE 55: QUCS simulation of the AGC decay period, 1s/div.

from full gain to maximum attenuation. **Figure 55** shows the decay period. The horizontal scale of this plot is 1s/div. The transition from maximum attenuation to full gain takes about 1s.

AGC Circuit Construction

The AGC unit is built on a strip of PCB. I used a pair of BC547 transistor in the 10.7MHz amplifier. The FET is a BF245C (Maplin N65AG). The 1μF capacitor is a Wimo plastic type. You can use a standard electrolytic instead provided that it has very low leakage.

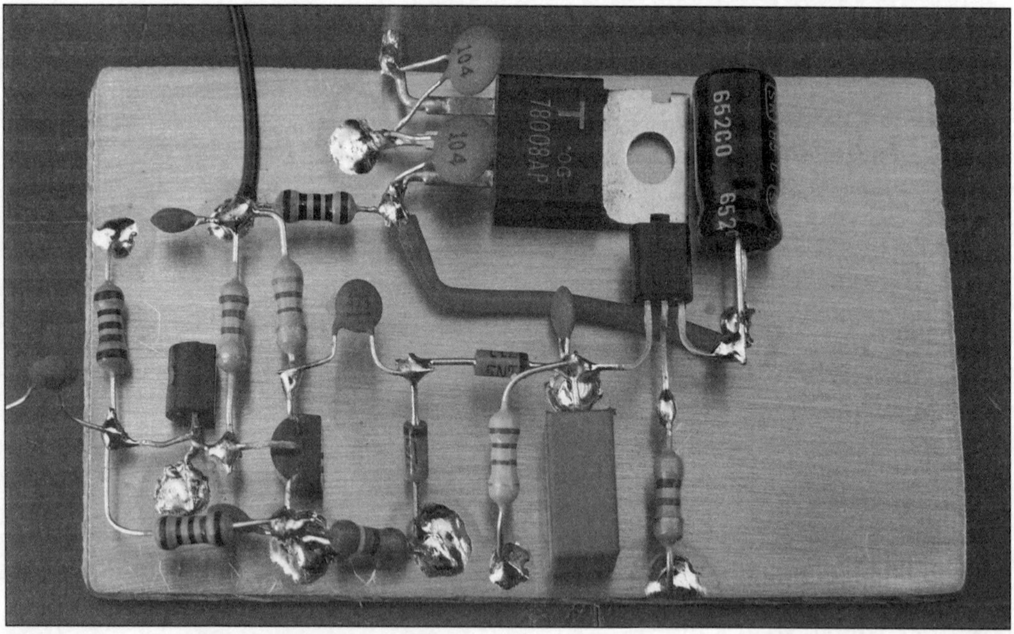

PHOTO 29: The finished AGC unit.

The S-meter is a 50μA SEW type. A 100μA or 200μA meter should also work in this circuit. **Photo 29** shows the finished AGC unit.

My own preference is for a very slow AGC decay. I don't like listening to rapid variations in background noise as the AGC operates. Other operators may prefer a faster AGC rate. **Figure 56** shows a simple way of providing a switchable fast/normal/slow AGC control.

Testing

The AGC unit should be installed in a position that is not too close to the input of the main IF. Connect the AGC output to the receiver AGC line. Set the meter-zero adjustment to the mid position and the meter deflection adjustment pot to maximum resistance (minimum deflection). Once the unit is powered up, set the meter zero adjust with no aerial connected to the receiver.

The 1k pot at the AGC amplifier input should be set for best dynamic behaviour of the AGC system. I found a value of 100Ω was about optimum. I removed this pot from the finished circuit and replaced it with a fixed 100Ω resistor. The meter deflection pot should be set for almost full scale on a very strong broadcast station. There is some interaction between the zero and deflection adjustments, so you might have to repeat the meter adjustments several times. If you have a calibrated signal generator, you can use it to calibrate the meter scale. Otherwise, the best that you can do is to calibrate the meter for S9 on a moderately strong signal. A switchable step attenuator can be used for relative calibration, ie S9+10dB, +20dB, +30dB...

The AGC has performed very well in listening tests. Even very strong local signals are well controlled without any evidence of distortion, popping clicking etc. As mentioned previously, I found that the receiver had far too much gain for use at LF/MF. The

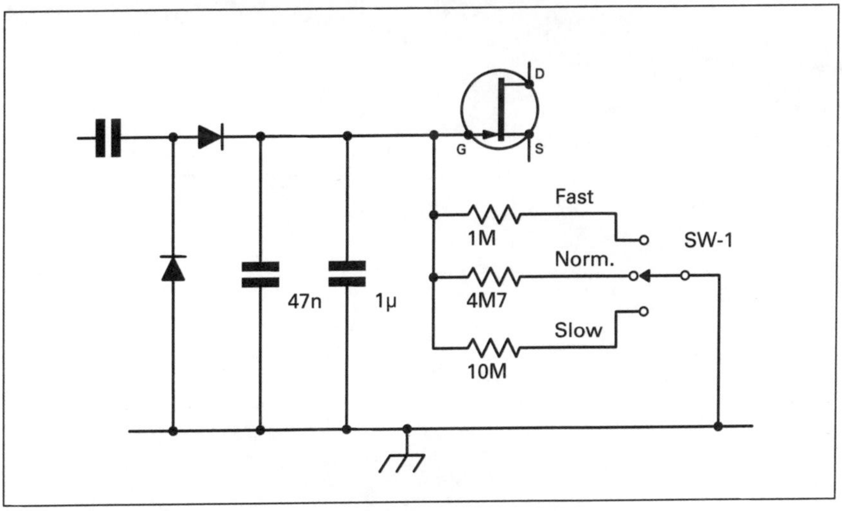

FIGURE 56: It's easy to add a switch for fast/normal/slow AGC.

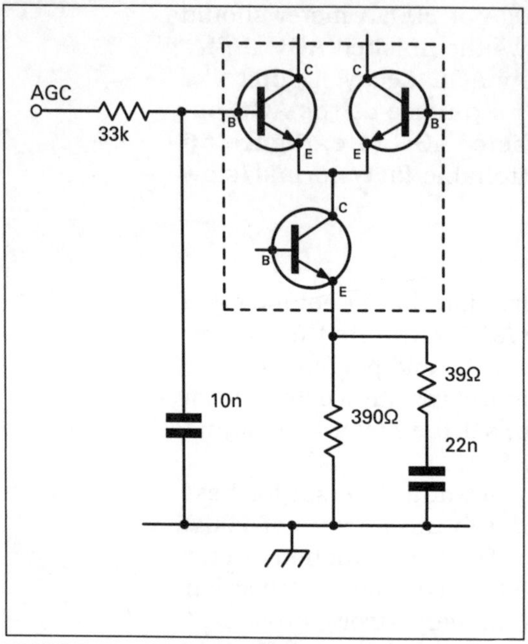

FIGURE 57: Modifications to the receiver IF stages to reduce the gain slightly.

total gain of the IF amplifiers is more than 100dB (20+40+40). The two audio stages have a combined gain of almost 80dB. Even when the losses of the various filters, passive mixers and attenuators are taken into account, the net receiver gain is around 150dB. This is enough gain to turn a sub-microvolt signal at the aerial into several volts at the speaker. I have made a couple of modifications to the IF amplifiers to reduce the gain to a more reasonable level. **Figure 57** shows the modifications to the IF stages. I have put a 39Ω resistor in series with the emitter bypass resistor of each IF amp. This reduces the gain of each stage to just under 30dB. I also increased the value of the resistors at the AGC input from 10k to 33k, which gives a slightly less sharp 'knee' in the AGC response curve.

[12] www.freepatentsonline.com/1342885.pdf
[13] Homebrew June 2006

8

Local Oscillators

I HAVE BEEN rooting through the EI9GQ junk box looking for suitable devices to use for the local oscillator section of the new HF transceiver. My old rig uses a band-switched VFO that was free running (too darn free!) in my original prototype. I later added a simple PIC based 'huff & puff' frequency stabiliser. Readers interested in the huff & puff technique should visit www.hanssummers.com where you will find a large collection of huff & puff articles including the original huff & puff article by Klaas Spaargaren, PA0KSB.

My first huff & puff stabiliser was a PIC implementation of the classic huff & puff circuit that is basically a 1 bit frequency counter. This configuration worked brilliantly for reception, but it was less than ideal for transmitting. The stabiliser 'locking points' were separated by about 10-20Hz. When the VFO was tuned to a new frequency, it would drift to the nearest locking point and then stay there. This scheme allowed me to tune to within 10Hz of any frequency on any HF band. The problems began when I started using the transmitter. Every time I switched from receive to transmit, the slight change in supply voltage and the mechanical shock of the relays caused a very slight frequency change of about 10Hz. This would normally be an insignificant change. However, it was just enough that it sometimes pulled the VFO frequency down to the next adjacent locking point. The receiver would remain locked on frequency for hours, but as soon as I started transmitting, the frequency would walk downwards in 10Hz steps.

My second attempt at building a PIC based VFO stabiliser was more successful. This used the PICs built-in counter/timer to measure the frequency of the VFO to a resolution of 24 bits. The single bit count of a simple huff & puff just gives a high/low indication relative to each locking point. The 24 bit counter gives an accurate frequency with a maximum value of 224 = 16,777,216Hz for a 1s gate time or 167.772MHz for a 100ms gate time. The adjacent locking point concept still applies, but the locking points are now 167MHz away from the locked frequency. This stabiliser was based on the PIC16C54 and it was published in Elektor **[15]**. I am currently using a similar stabiliser based on the reprogrammable PIC16C84 (and F84) **[16]**. This unit displays the VFO frequency to a resolution of 10Hz and indicates the lock status. Frequency correction is applied to a varicap diode in the VFO. This allows the VFO frequency to be slowly steered by about ±2kHz. It cannot correct more extreme drift, nor can it track rapid changes in frequency. In short: it can't make a good VFO out of a really bad one but it can make a good VFO even better. As long as the unit is locked, long term drift is the same as the PIC crystal reference oscillator.

The simple frequency counter project from the October 2006 Homebrew would make a good basis for a huff & puff stabiliser. As there are several I/O port pins left unused on the PIC. It would be a fairly simple matter to add the extra components to make a LPF/integrator for the error correction pulses from the PIC. It should also be possible to use the 10 bit PWM output of the PIC to provide the frequency correction pulses.

The other obvious choice for the local oscillator is one of the newer programmable clock generator chips like the Silicon Labs Si570. These PLL-based devices can produce a clock signal at frequencies from around 10MHz to UHF. The output frequency can be programmed by a PIC or other microcontroller via an I^2C serial interface. Amateurs using these devices as receiver local oscillators have reported that they produce low phase noise and low spurious signals. This makes them a viable alternative to the DDS in this application. Hans Summers' website (mentioned earlier) has a detailed DDS vs Si570 comparison.

My junk box search produced several DDS devices: a couple of AD9851s as used in previous projects, a couple of AD9852s and a single AD9951. The AD9851 has given very good service in some of our previous projects. It is a reasonably good performer, but the level of spurious outputs is a bit too high for use in a high performance receiver. The spurs are generally at or below the noise floor on the low frequency bands. On 10m, or when I am transverting to VHF where band noise is at a low level, DDS spurs often appear above the noise.

The AD9852 is a 300MHz DDS with a 10 bit DAC. This offers similar performance to the AD9851 except that the maximum output frequency will be a bit higher and spurs may be a bit lower because of the higher maximum clock frequency.

The 400MHz AD9951 and the 1GHz AD9912 are newer devices, with 14 bit DACs. The improved design, extra four DAC bits and higher clock frequency results in a significant reduction in DDS spurs. These 14 bit DDS chips are used in several of the more recent commercially made HF transceivers.

A hybrid DDS/PLL of the type used in the Homebrew June/ July 2010 VHF synthesiser project would probably offer the best compromise between phase noise and spurs. Unfortunately, a band switched version covering all HF bands would be horribly complicated. It would seem that the AD9951 is the best of the available choices. The choice is made easier by the fact that I have one gathering dust in the shack.

DDS Clock

The long and short term phase and frequency stability of a DDS depend on the quality of the clock signal. This means that the clock oscillator is a key component in the DDS system. Because I have become accustomed to the extremely good frequency accuracy and stability of the AD9851 DDS I am using in my current transceiver, I will expect to achieve similar performance from the new unit. OCXOs are now widely available on internet auction sites like eBay. The most sought-after (and therefore most expensive) units are 10MHz OCXOs. Oscillators with odd frequencies like 101MHz are not in great demand and can be bought very cheaply. As the DDS clock frequency can be any arbitrary value as long as it is several times the maximum output frequency, these OCXOs would make a very good DDS clock. The 245MHz Vectron units sell for around £25. These would be ideal for our application because they could be used directly without any need for frequency multiplication.

I have a couple of spare 100MHz OCXO that could be used for this project. The maximum output frequency of my DDS will be around 52MHz – 10.7MHz = 41.3MHz. I would like to allow for a higher output frequency in case I want to cover the 4m band at a later time. 70.7 – 10.7 = 60MHz or thereabouts. The old rule of thumb that the clock should be more than three times the maximum output frequency suggests that 180MHz is the absolute lower limit for the clock frequency. I would prefer to use an even higher clock frequency so that I can keep spur and alias signals to a minimum. The AD9951 has a built-in PLL based refclock multiplier that can be programmed to multiply the clock frequency by various values ranging from 4x to 20x. This means I could get a 400MHz clock by using the 100MHz OCXO and 4x multiplier or a 200MHz clock from a 10MHz OCXO and 20x multiplier. Using the internalmultiplier comes at the cost of increased close-in phase noise on the DDS output signal. Before I try using this approach, I would prefer to have a go at building a VHF DDS refclock based on a 100MHz OCXO and a frequency doubler or tripler. Frequency multiplication in the external refclock will also cause some phase noise increase at a theoretical rate of 6dB per octave, but this will probably still be cleaner than the internal PLL based refclock multiplier.

The complete OCXO and doubler circuit is shown **Figure 58**. The diagram indicates a typical pin configuration for an encapsulated OCXO; specific voltages and pin numbers are not shown because they tend to vary depending on the type of OCXO used. Most OCXOs have an EFC (electronic frequency control) input that can be used for fine tuning the output frequency or phase locking to an even more stable reference like a rubidium standard or GPSDO.

The output from the OCXO is buffered by an emitter follower. This may not be necessary if the oscillator is designed to produce 1V or more into a 50Ω load. However, it is always a good idea to at least put a coupling capacitor between the OCXO output pin and any external circuit. This will prevent DC voltage in the external circuit from damaging any transformer or low value resistor that may be in the OCXO's internal buffer amplifier.

T1 in the frequency doubler is 6 turns of enamelled copper, trifilar wound on a FT37-61 ferrite toroid. L1 and L2 are each 4 turns of 1mm copper wire (TV coax core) wound wide spaced on a 5mm former. The output tap on L2 is 1 turn from the ground end. The 20pF trimmers are Philips green types. 20pF UHF piston trimmers would be preferable if you have some available. **Photo 30** shows the assembled oscillator/doubler unit.

I am sometimes foolish enough to expect such a simple circuit to work properly at the first attempt. Initial testing was a complete failure. There was plenty of output at 400MHz and almost none at 200MHz. The 400MHz output was almost 30dB stronger on the spectrum analyser. I suppose I should have just cut my losses, declared the job finished and used the 400MHz signal as the DDS clock. It is not that easy to debug VHF/UHF circuits because my oscilloscope bandwidth is limited to 60MHz. I eventually came to the conclusion

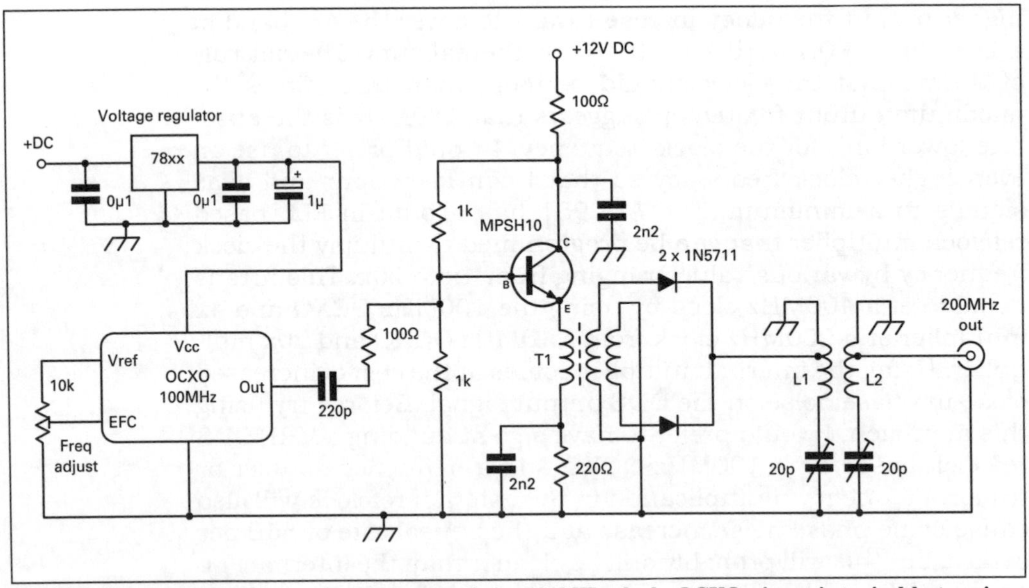

FIGURE 58: 100MHz OCXO and doubler circuit for 200MHz clock. OCXO pinout is typical but varies depending on device type.

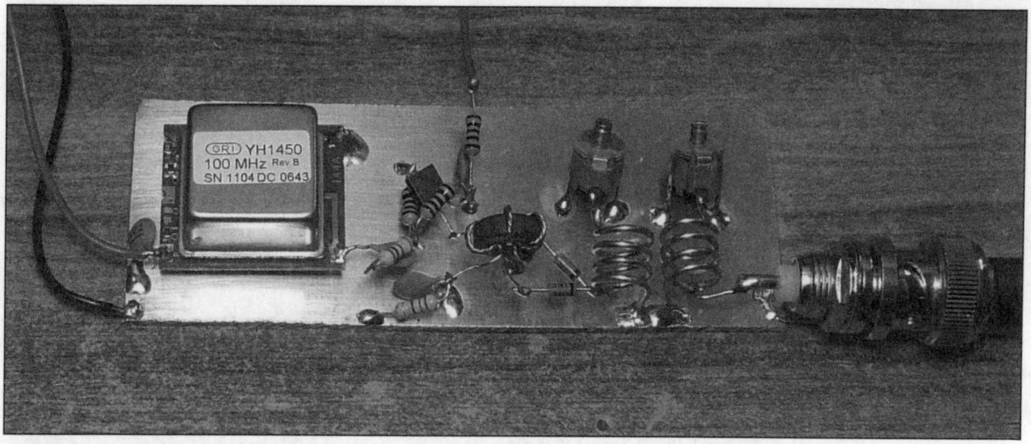

PHOTO 30: 200MHz reference clock consisting of a 100MHz OCXO module (left) and doubler circuit.

that the relatively high output from the OCXO was overdriving the MPSH10 buffer to the point where it makes a surprisingly good frequency doubler. The resulting 200MHz is doubled again by the diodes. The problem was cured by placing a 100Ω resistor in series with the buffer input, as shown in the schematic. The circuit now produces a nice 200MHz signal with the 100MHz fundamental and 300MHz harmonic very well suppressed.

Figure 59 shows a frequency tripler that produces an output at 300MHz. L1/L2 are 3 turns of 1mm wire on a 5mm former. The L2 output tap is at 0.75T above ground. Output from this circuit is slightly stronger than from the doubler circuit, at just under +3dBm. The 200MHz and 400MHz outputs are only 30-40dB below

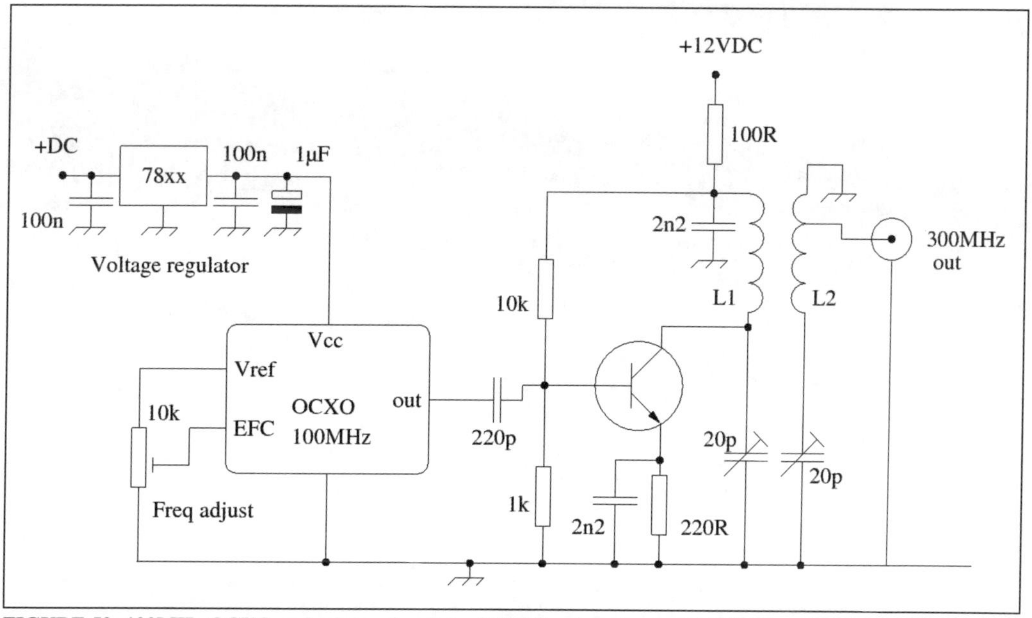

FIGURE 59: 100MHz OCXO and tripler circuit for 300MHz clock; typical OCXO pinout.

the 300MHz signal. I will add a third resonator to the output filter to see if this will improve the spectral purity of the output signal. This would then be an ideal clock oscillator for the DDS.

[14] A2087 (CV2171) data sheet, Marconi, 1959
[15] Elektor Electronics, February 1998, Frequency display and VFO stabiliser, E Skelton, EI9GQ
[16] http://homepage.eircom.net/~ei9gq/stab.html

9

A DDS VFO

A BETTER OSCILLATOR - So far, I have been using the AD9851 direct digital synthesis (DDS) VFO [17] as the local oscillator for the HF transceiver project. This has performed reasonably well, but there is definitely room for improvement.

The newer generation of DDS devices offer significantly improved performance when compared to older devices like the AD9835, AD9850 and AD9851. The AD9951 and AD9912 are widely used by builders of high performance radio equipment. These devices have a specified maximum clock speed of 400MHz and 1GHz respectively. A high clock rate allows a high maximum output frequency and reduced spurs at lower output frequencies.

The newer devices have a 14 bit digital to analogue converter (DAC) instead of the 10 or 12 bit DAC used in older DDS chips. Each additional DAC bit adds 6dB to the DAC output range. The newer chips also use a much lower supply voltage. The DDS core of the AD9951 runs at 1.8V. The parts of the chip associated with digital input/output (I/O cells) can be run at a higher supply voltage of 3.3V. This allows for direct interfacing with 3.3V logic. Some of the AD9951 digital input pins are 5V tolerant. This is very convenient when the DDS is controlled by a 5V microcontroller like the Pinguino PIC board **[18]**. Power consumption in CMOS circuits is proportional to clock frequency. Using a low supply voltage allows for very high clock frequencies while keeping power consumption within acceptable limits. Advanced DDS devices have much in common with computer CPUs. The CPU in my desktop PC has a core supply voltage of 1.4V and a maximum clock frequency of 3.2GHz. This is dynamically reduced to 1.1V and 1GHz when energy saving mode is active. The latest CPUs run at even lower voltages and higher clock speeds. Hopefully, this state-of-the-art technology will make

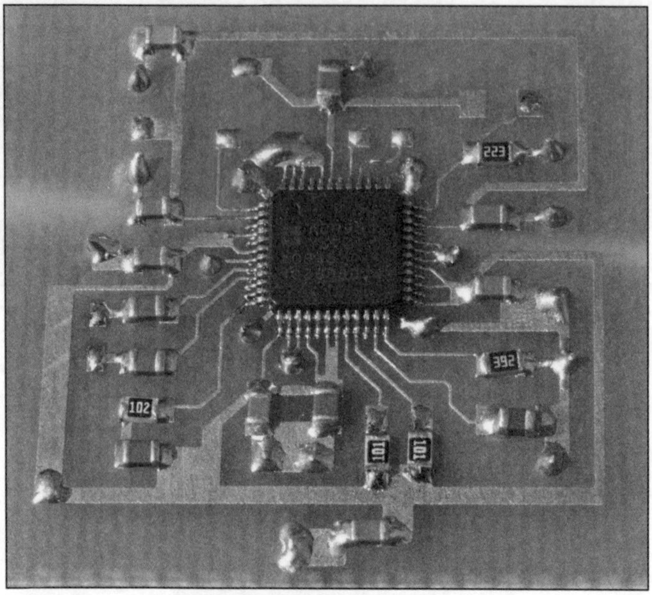

PHOTO 31: The completed AD9951 DDS board.

its way into the next generation of DDS ICs. I hope they come in a homebrew-friendly package...

I have chosen the AD9951 for the new DDS project. With a clock frequency in the 300-400MHz range, this chip should be capable of providing a clean local oscillator signal for all of the HF bands and the two lowest VHF bands. As I intend to use a high quality 100MHz OCXO (Homebrew, August 2011) for the refclock, I have four options for refclock: 100MHz direct from the OCXO, 200MHz using a frequency doubler, 300MHz using a frequency tripler and 400MHz using the 100MHz OCXO with the DDS internal refclock multiplier set for 4x. As the first two options result in a refclock frequency that is too low for my application, 300MHz (tripler) and 400MHz (refclock multiplier) are the only viable options. The AD9951 also has a built-in crystal oscillator that can be used with the internal refclock multiplier. My PCB layout does not support this option.

The DDS IC and a few critical components are mounted on a small PCB (**Photo 31**), while the power supply voltage regulators are mounted on a separate board. Other peripheral circuits like the output filter are not included on the DDS carrier PCB. This modular approach might seem untidy compared to a single PCB solution, however it does offer some advantages. Keeping the DDS separate from electrically noisy circuits like the PIC micro can help to reduce unwanted noise in the DAC output. The modular DDS also tends to be more flexible. Keeping the DDS PCB small and relatively simple makes it easier to build – and reduces the amount of rework required if things go horribly wrong. There is always a good chance that the first version of a PCB will end up in the bin.

Figure 60 shows the AD9951 pinout. In common with other DDS devices made by Analog Devices, separate power pins are used for the digital and analogue parts of the circuit. Most constructors will be quite familiar with the idea of using separate power supplies for different parts of a circuit. It is quite usual for two separate circuit blocks to have separate power supplies (or well decoupled supplies from the same source) for the positive supply only. In most cases, the negative supply is via a negative supply rail or a ground plane that is common to both circuits. The AD9951 is designed to use separate power supplies for both the positive and negative supply pins. This means that even in the usual negative-ground configuration, separate ground planes are used for digital (DGND) and analogue (AGND) ground. In practice, this means that there is no direct connection between the DGND and AGND planes of the PCB. However, the two grounds may well be connected at some remote point such as the equipment chassis or the negative terminal of a battery.

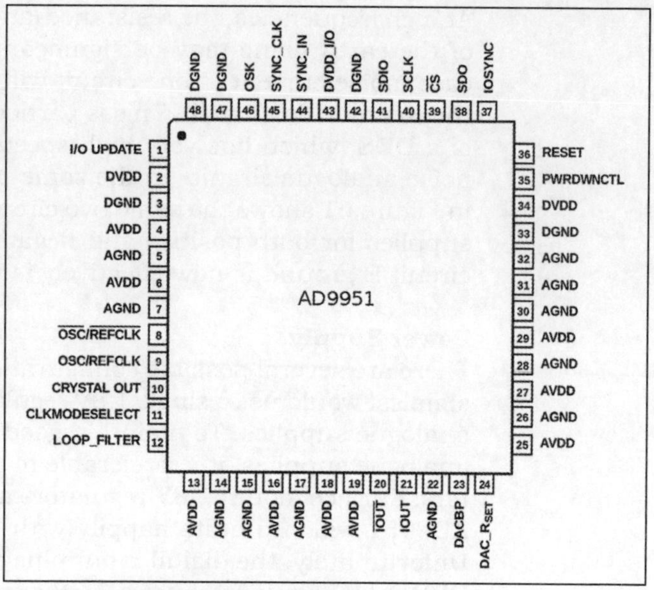

FIGURE 60: Pinout of the AD9951 DDS IC.

The simple schematic at the top of **Figure 61** shows a typical negative-ground configuration for a pair of circuit blocks. The ground/chassis/negative connection is common to both circuits. This arrangement works perfectly if the complex impedance of the ground plane is zero.

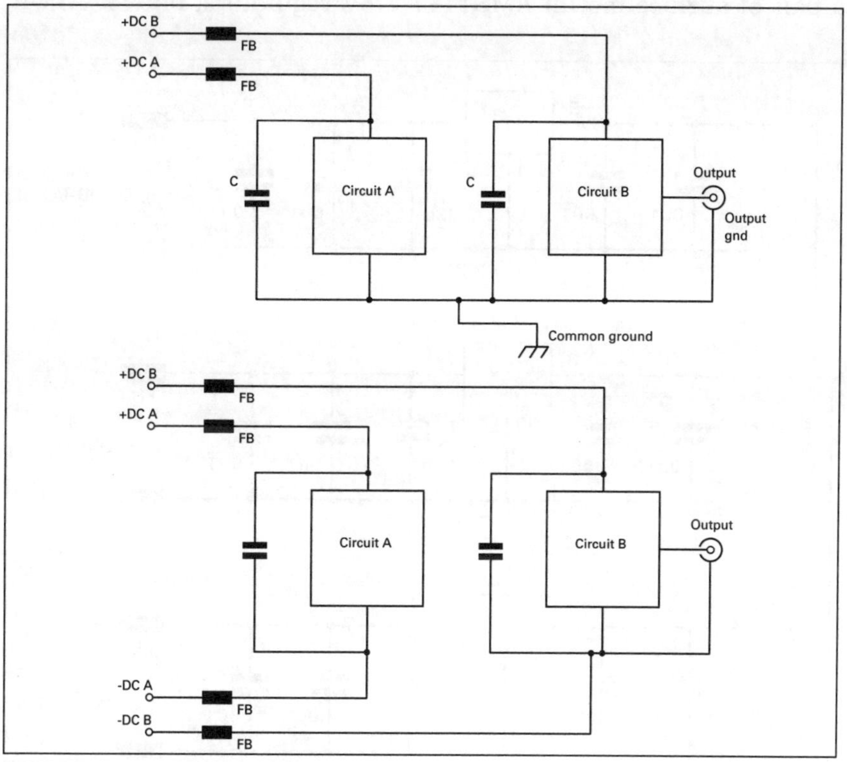

FIGURE 61: Grounding strategies (see text).
Top – common grounds.
Bottom – separate grounds.

At high frequencies, the resistance (and particularly the inductance) of the earth plane may be significantly greater than zero. In this case, noise current in one circuit will produce noise voltage across the common earth path. This is particularly troublesome in the case of a DDS, which has very high speed digital circuits and very low noise analogue circuits in the same package. The lower schematic in Figure 61 shows the same two circuit blocks with separate power supplies for both positive and negative power rails. Note that the circuit B ground is now effectively isolated from circuit A.

Power Supply
There are several possible configurations for the power supply. The simplest would use a single 1.8V regulator to provide both digital and analogue supplies. To maintain good isolation between digital and analogue supplies, it is preferable to use two separate 1.8V regulators. Surface mount 1.8V regulators are small and quite cheap. The AD9951 will run quite happily with just 1.8V on all supply pins. Unfortunately, the digital input pins are not 5V tolerant unless the DVDD_I/O pin is fed from a 3.3V supply. This means that an ideal power supply configuration will need at least three separate supplies, each with its own voltage regulator IC.

The resulting power supply circuit is shown in **Figure 62**. I used a pair of surface mount AMS1117 1.8V regulators, one for AVDD

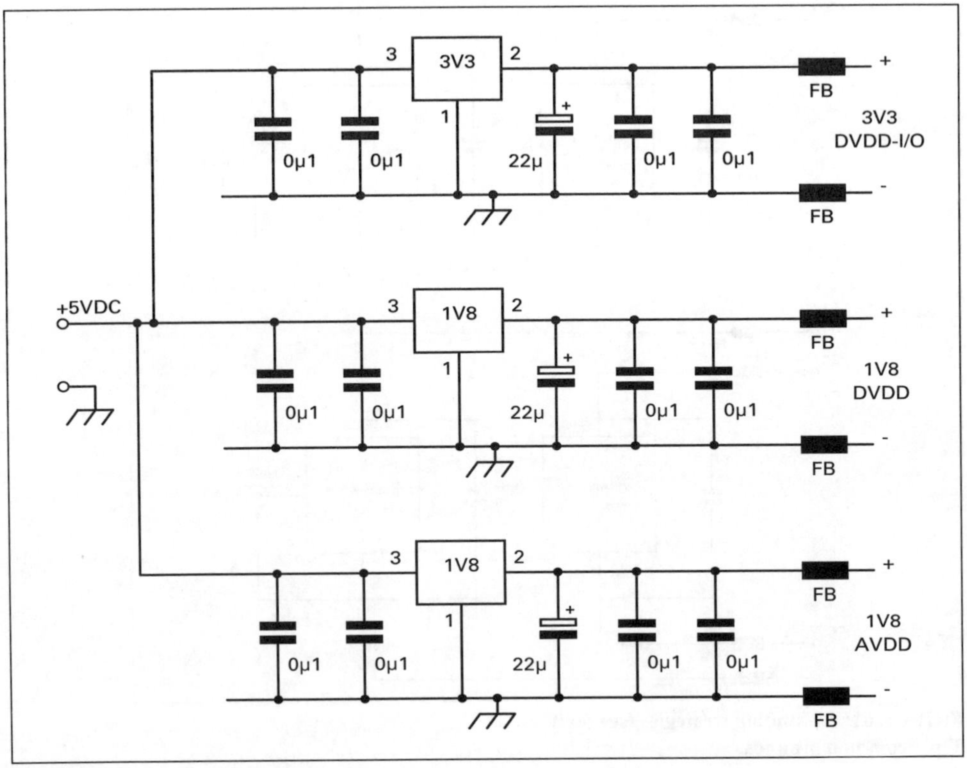

FIGURE 62: Voltage regulator circuit.

and one for DVDD. As I didn't have a surface mount 3.3V regulator, I used a larger TO¬-220 size regulator instead. The 3.3V DVDD_I/O regulator is an LD33CV. All three regulators use the same pin configuration as the LM1117: pin 1 = gnd, pin 2 = output, pin 3 = input. Note that the heatsink tab is internally connected to pin 2 (output). This means that the heatsink tab must not be connected to chassis/ground. This pin configuration is not universal; if you use a different regulator, check the data sheet for pinout details. You should also measure the three output voltages carefully before you connect the supply lines to the DDS board. 5V or even 3.3V on a 1.8V supply pin would probably have fatal consequences for the DDS IC.

The regulator unit is made from two strips of single sided PCB. The narrower top strip is superglued to the bottom strip. The assembled unit is shown in **Photo 32**. The copper foil of the top strip is cut into three sections for the three supplies. From left to right they are 3.3V DVDD (for I/O), 1.8V DVDD, 1.8V AVDD. You may just be able to see the surface mount 0.1µF capacitors to ground from pins 2 and 3 of the 1.8V regulators. A ferrite bead is placed on all six wires to the DDS board. The negative lines for both digital supplies are connected directly to the DGND plane on the back of the DDS board. The negative wire from the 1.8V analogue supply is connected to the AGND plane on the back of the DDS board. The positive DVDD, DVDD_I/O and AVDD are connected to their respective connection points on the DDS board. Take great care to ensure you don't connect the 3.3V supply to the wrong place. The output voltages are taken from the three rectangular tabs on the top strip of the PCB. Note that my unit has a potential hazard because I used a 3mm metal screw and nut to secure the regulator to the PCB foil. The other side of this screw should be isolated from ground. This issue won't arise if you use three surface mount regulators or if you use a 3mm nylon screw instead.

PHOTO 32: Complete voltage regulator board.

The AD9951 DDS

The AD9951 data sheet **[19]** contains a lot of useful information. It is a fairly complicated device and some of the information is a bit terse; I found it difficult to follow. The excellent DDS projects by G3XJP **[20]** and G6ALU **[21]** provided a lot of help and inspiration. While I was working on the PCB layout, there were several occasions when I was unsure of what I should do with an unused pin. Tie it high, tie it low or leave it unconnected? Referencing these known good projects provided a quick and easy answer. For example: the PWRDWNCTL pin. Both of the above projects have it tied to DGND. I would probably have figured this out for myself eventually, but it was nice to have a proven reference. Many thanks to both authors.

Good grounding and power supply decoupling is essential in a DDS. The manufacturer's evaluation and development boards are usually multi-layer boards with plated through vias (inter-layer connections). Kitchen table boards are generally limited to two layers and hand soldered wire vias. This limits the number of vias because hand soldered ones are big and ugly. It also means that vias cannot be placed under ICs. This hasn't been a big limitation in the past, but this is changing now that BGA (ball grid array) and similar packages are becoming more common. This type of chip has all connections under the IC.

The analogue ground is a particularly important part of the DDS circuit. This is the ground plane for the DAC and other sensitive analogue circuits. As we are concerned with suppressing all noise and spurious signals, even at levels of 80-100dB below the wanted output, we must take great care to keep noise out of the AGND and AVDD circuits. In addition to the AGND pins listed in Figure 60, the conductive pad on the bottom of the chip must be connected directly to AGND. The only practical way of doing this on a home-made PCB is to drill a 3mm hole through the PCB at the exact centre of the IC

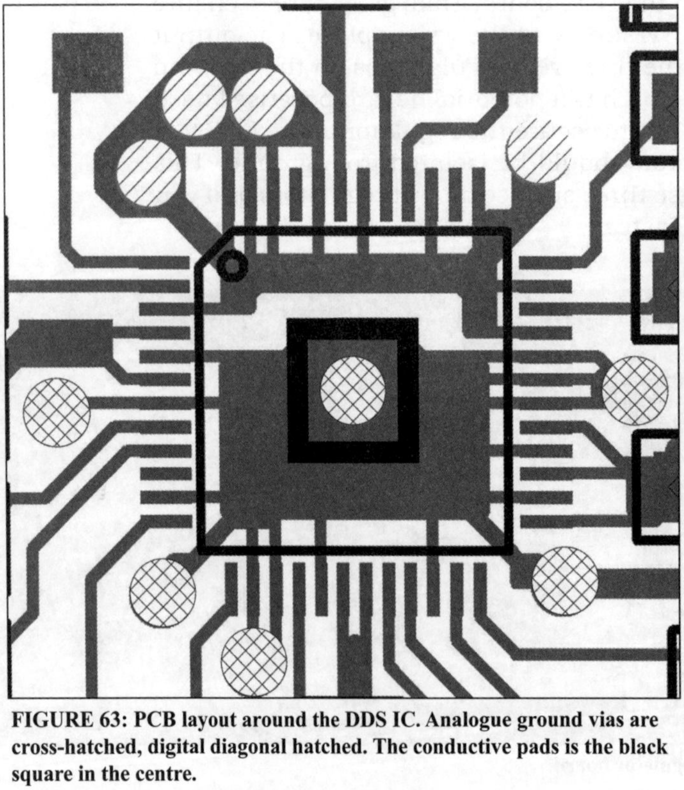

FIGURE 63: PCB layout around the DDS IC. Analogue ground vias are cross-hatched, digital diagonal hatched. The conductive pads is the black square in the centre.

pads. A short length of copper braid can be used to make a very short and direct connection between the top and bottom AGND copper foil and the IC pad. The G3XJP DDS article **[20]** gives a very clear and detailed description of this process.

Ideally, every ground pin would have a direct connection to the ground foil on both sides of the board and every power supply pin would have several decoupling capacitors. This is not quite possible because of the size of the components. I used type 805 surface mount capacitors and resistors. These are quite small compared to through-hole components, but smaller SM components are available. The 805 package is a reasonable compromise between size and solderability. A total of nine decoupling capacitors are used around the DDS IC, four on the digital side and five for AVDD. **Figure 63** shows the PCB layout in the immediate vicinity of the DDS IC. Analogue ground vias are shown as a cross hatch, digital ground vias are shown as diagonal hatch. The outline of the conductive pad is shown as the black square in the centre The AGND pins are connected directly to the top AGND plane and to the main AGND plane on the bottom of the PCB using the conductive pad and five additional wire vias. The digital ground is similarly connected top and bottom using four vias.

PCB

The artwork for the DDS PCB is available in native PCB and in Postscript format. The current version is at http://homepage.eircom.net/ ~ei9gq/art.html.

Note that the track pattern is reversed so that the toner will be on the PCB side of the film. Check this carefully before exposing the PCB. You don't want to make a nice PCB that is a perfect mirror image of mine! I used my old HP laser printer to print the track pattern on Staedtler OHP film. This printer has seen better days and I suspect it is overdue for a new toner cartridge. I find I can only get sufficient contrast (opacity of the track pattern) if I use two stacked sheets of film. To save OHP film, my artwork has two copies of the track pattern. Simply cut the OHP film in two and you have two copies of the artwork.

Because of the very small size of the IC pads, accurate alignment of the stacked sheets is quite difficult. I used a strip of masking tape to fix the first sheet of film to the glass exposure frame and a second strip to hold the second film in alignment. I spent about ten minutes with a good magnifying glass before I was happy with the alignment. The PCB is double-sided photo sensitive fibreglass board (Maplin FA61 or similar). I used a 500W halogen lamp to expose the board. The process is described in detail later see Appendix 1. After exposure, the board was developed using a sodium metasilicate solution (Maplin AP01B), rinsed and then etched in a ferric chloride bath. After a final rinse, the etch resist was removed using fine steel wool. The bottom of the board is not etched.

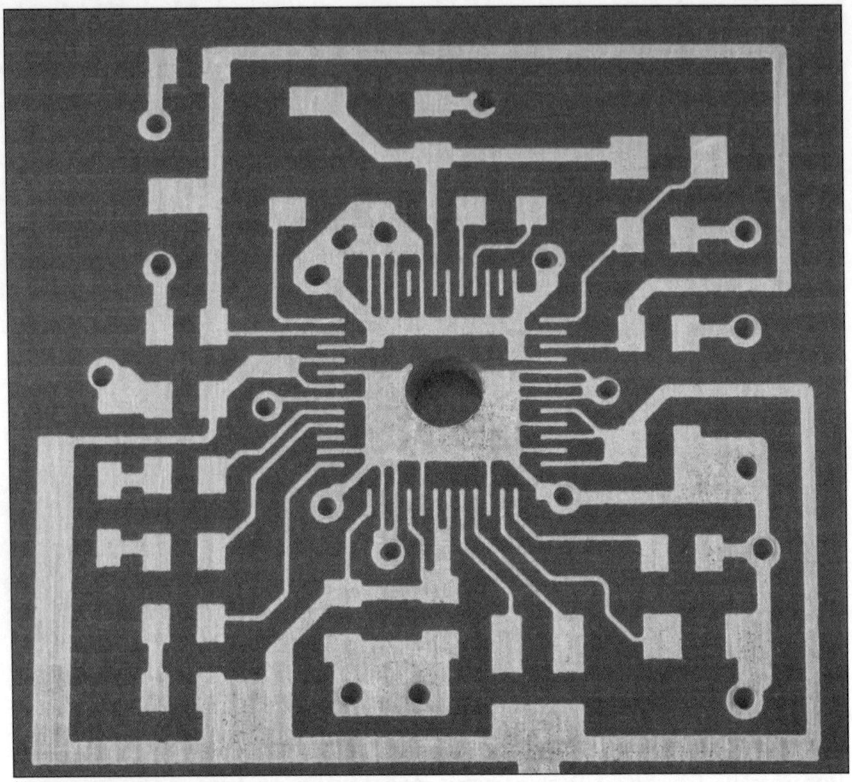

PHOTO 33: The DDS PCB copper side, fully drilled..

The gap in the pads at each corner of the 48 pin IC provides space for a reasonably wide PCB track. This space was exploited at all four corners to make connections between the top and bottom ground foils. For my first attempt, I used the standard 'signal' via size in PCB **[22]** for most of the vias. This turned out to be a rather unfortunate mistake. After the first batch of two boards were etched and cleaned, I discovered that this via size is effectively undrillable. Any drill larger than 0.6mm would completely strip the annulus (copper ring) from around the via hole. Drill sizes of 0.5mm or less are hard to find and extremely easy to break. The revised layout uses larger vias, not the next size up 'power' via, which is too big, but the signal via with the size incremented by three steps. For PCB users, tap the 's' key three times while the cursor is over the via.

I used a 0.6mm drill for the via holes. I managed to drill them all without breaking a drill. A 3mm hole was drilled through the via marker in the centre of the IC pads (see later). **Photo 33** shows the topside of the board after drilling.

To separate the AGND and DGND on the bottom of the board, the foil is simply cut in two using a hacksaw at a strategic point under the DDS IC. Spend some time identifying the AGND and DGND via holes. You don't want to cut in the wrong place. The rough edges of the cut were removed using a carpet knife and the gap was deliberately widened to reduce capacitive coupling between AGND and DGND. **Photo 34** shows the bottom of the board. Use the photograph and the via holes as a guide.

The wire vias are bare 0.4mm copper. This was stripped from standard 4 core telephone cable. The wire is threaded through the via holes from the bottom side, bent through 90 degrees and then soldered to the bottom foil. Use as much solder and heat as you like. You should be more careful with the top side. Snip the wire at about 1mm above the board and solder the via using a small pointed tip and very fine solder. **Photo 35** shows some of my AGND vias. The outer two are good. The middle one was a bit grotty, so it was resoldered after the photo was taken.

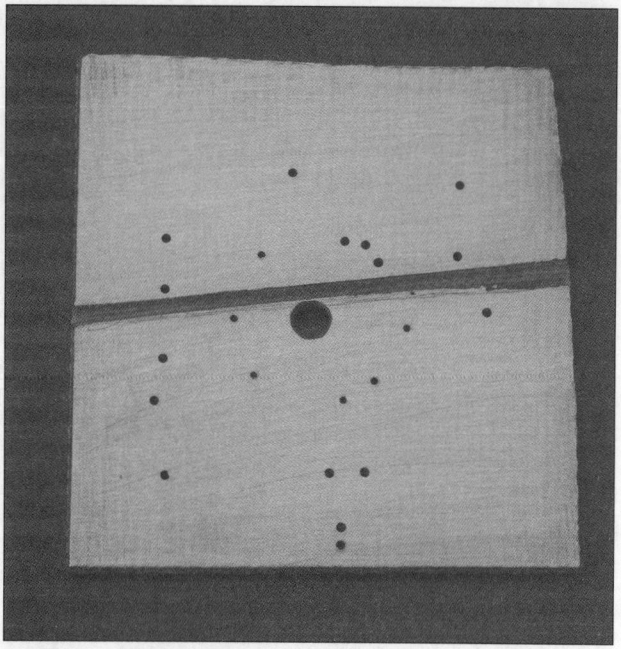

PHOTO 34: Bottom of the board, showing the wide cut that separates AGND & DGND..

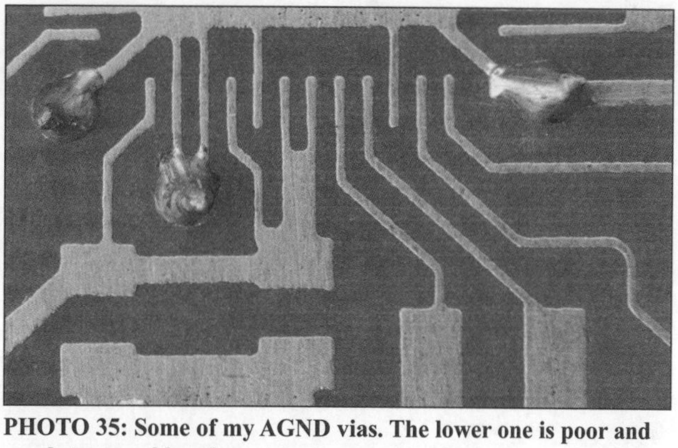

PHOTO 35: Some of my AGND vias. The lower one is poor and was later re-soldered.

The capacitors and resistors were hand soldered using a small pointed tip and very fine solder. The components were held in place using a wooden cocktail stick while the first joint was made.

Mounting the DDS IC was relatively easy. The edge of the AGND side of the 3mm hole was tinned using a small amount of solder. This will help the IC pad to make a good connection to the top foil. Excess solder was carefully removed using desoldering braid. Any small blob of solder here will prevent the IC from seating properly on its pads. Excess solder could also form a bridge between AGND and DGND. I spent several minutes checking and re-checking the pad alignment before tacking the corner pins in place using a small pointed tip. Soldering the chip in place took just a few seconds, using the drag soldering technique. Each row of pins was just enough solder and heat to make a connection to the top AGND foil and the IC pad. The braid was then soldered to the bottom AGND foil.

The finished board is shown in Photo 31. After construction and carefully checking the DDS board and the regulator unit, the DDS board was wired into my DDS test rig. The serial clock, data and update lines were tied to DGND through 22k resistors, a 100MHz clock was connected to the refclock input and the unit was cautiously powered up by connecting a 5VDC supply to the input of the regulator unit. The chip ran lukewarm, but not hot, and all three regulators ran cool.

[17] Homebrew, March 2010 and November 2009
[18] see [17] and http://homepage.eircom.net/ ~ei9gq/picboard.html
[19] AD9951 data sheet: see www.analog.com
[20] Google "DDS-AD9951 for STAR"
[21] AD9951 DDS board: G6ALU – http://tinyurl.com/5todx3u
[22] http://pcb.gpleda.org

10

Continuing the AD9951 DDS Project

THE PCB ARTWORK is available at **[23]**. **Figure 64** (overleaf) shows the DDS schematic. This covers the DDS PCB only. The power supply voltage regulators, DAC output transformer (if required) and LPF are not included. I have followed the usual practice of placing a choke (inductor) between the AGND and DGND copper ground planes on the back of the PCB. This ensures that the two independent ground systems are at the same DC potential, even in the event of a failure in another part of the system. The choke consists of a ferrite bead (FB) threaded on a single wire as shown in **Photo 36**. This relatively large bead gives a choke inductance of 1.2μH. If a large FB is not available, you can stack two or more smaller beads instead.

There are a few differences between the AD9951 and the AD9851 used in some of our previous DDS projects. The refclock input to the AD9951 is via a pair of balanced complimentary clock inputs instead of the single-ended refclock input of the older DDS ICs. The DAC output of the AD9951 is referenced to AVDD and not to AGND. Serial data communication between the PIC controller and the DDS is essentially the same as with the older devices, although the DDS pin names are slightly different.

PHOTO 36: A ferrite bead on a wire link provides a DC link but AC isolation between the analogue and digital grounds.

The DDS-carrier PCB component overlay is

shown in **Figure 65**. The light gray line shows the point where the ground plane is cut in half on the bottom side of the PCB. As I am using a single-ended refclock source, the clock signal is applied to the OSC/REFCLOCK input only. The open end of the complimentary OSC/REFCLOCK input coupling capacitor is tied to AVDD using a short wire link. This is shown in diagonal hatching on the left side of the overlay. To allow maximum flexibility, I have not included conventional PCB pads for the DAC output transformer. Instead, there are three generic pads connected to IOUT, IOUT and AVDD. These pads can be used to mount the DAC output resistors and/or DAC output transformer. The connection points are marked with diagonal hatching on the overlay. The DAC bypass capacitor (pin 23) is a non-polarised 1µF SM capacitor and not a polarised electrolytic as suggested by the silk screen layer.

Clock Source

Using an OCXO as the refclock source guarantees that the DDS will have excellent medium to long term frequency stability. For our purposes, medium to long term means seconds, minutes, hours and

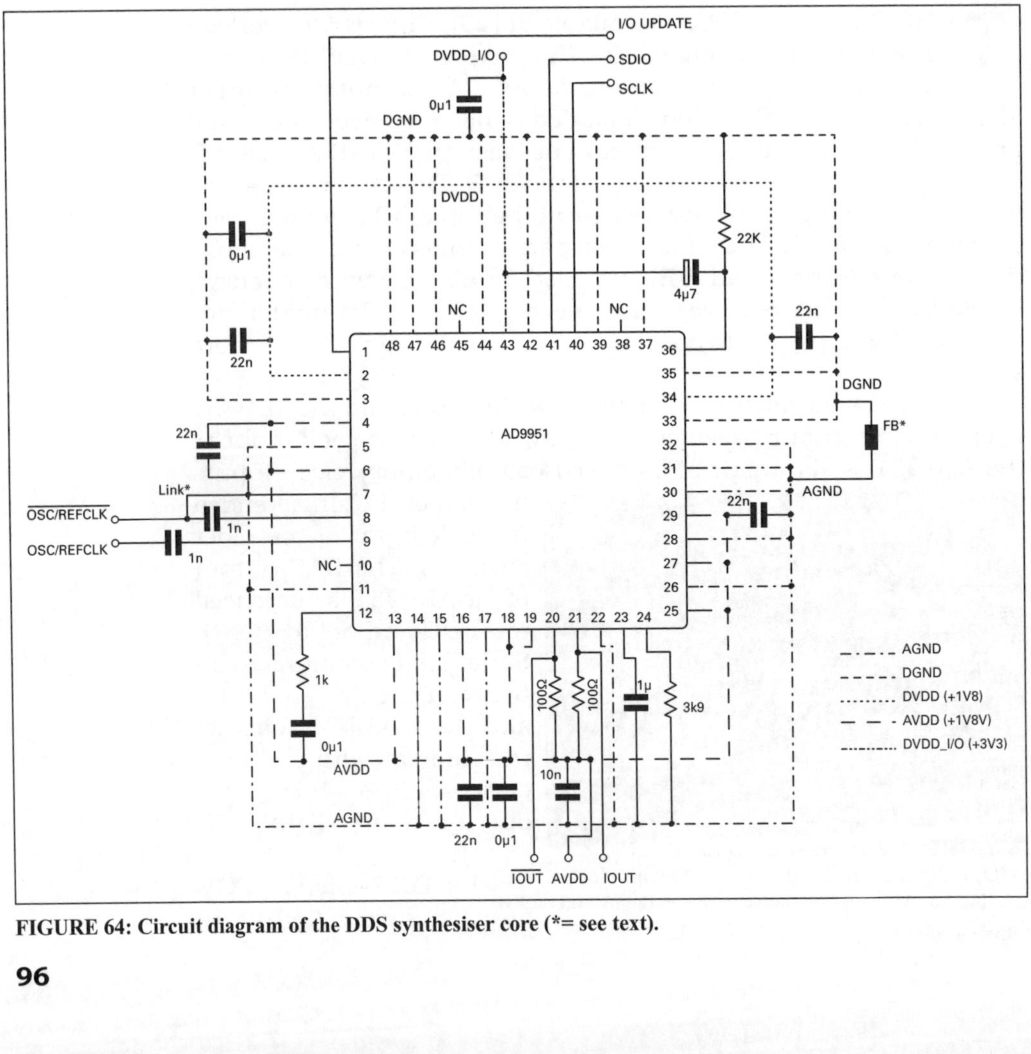

FIGURE 64: Circuit diagram of the DDS synthesiser core (*= see text).

days. Shorter term phase variations (or 'jitter') on the refclock signal will lead to unwanted phase noise at frequencies close to the refclock frequency. A well designed crystal oscillator as found in most OCXOs can produce a very clean signal which is relatively free of phase noise. A typical OCXO will produce phase noise at a level of around -140dBc/Hz at 1kHz from the carrier frequency. Some OCXOs that are specifically designed for very low phase noise can achieve -160dBc/Hz at 1kHz. Such an oscillator would make an ideal refclock for a DDS.

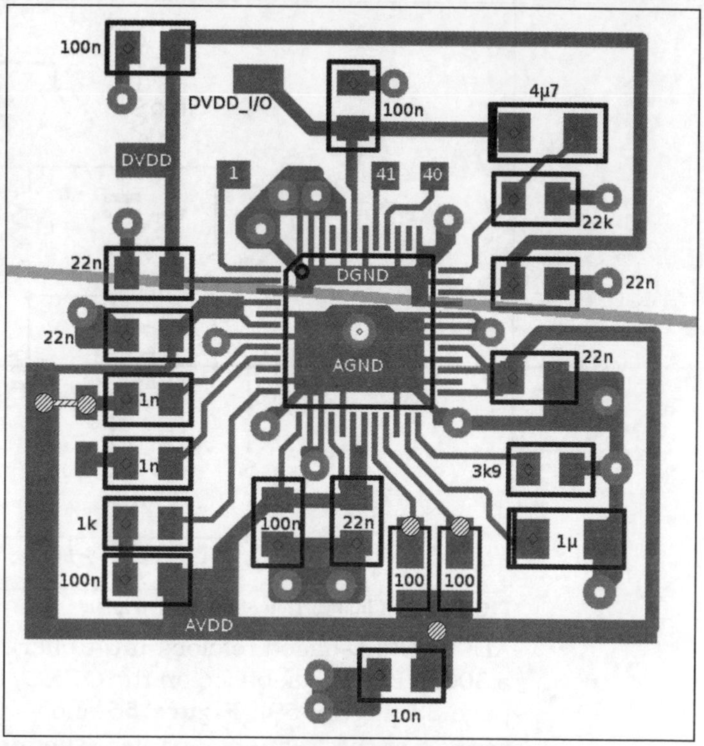

FIGURE 65: Component overlay for the DDS-carrier PCB.

In an ideal world, a DDS would act as a perfect frequency divider, which would reduce phase noise at the theoretical rate of 20LOG(N)dB, where N is the frequency division ratio. This would result in an output signal which always had less phase noise than the refclock. Unfortunately, a real world DDS is far from perfect. There are many sources of phase and amplitude noise in the DDS and its associated circuits. The DDS and its output DAC have limited precision, some distortion due to imperfections in the DAC hardware, noise within the analogue section of the DDS, noise leakage from the noisy digital section of the DDS, noise leakage from other sources like the PIC controller, noise from the voltage regulators on the power supply rails...

Some of this noise will appear as random phase and amplitude variations of the DDS output signal. The AD9951 data sheet specifies residual phase noise of typically -132dBc/Hz at 1kHz for a 40MHz output signal when the refclock multiplier is disabled. When the refclock multiplier is set for 4X, phase noise increases to -115dBc/Hz. Setting the refclock multiplier to the maximum value of 20x increases phase noise to -105dB/Hz. The 'Residual Phase Noise' specification is noise generated by the DDS after phase noise from external sources like refclock jitter and power supply noise is excluded.

For the initial testing of the DDS, I used a Greenray 100MHz OCXO with a specified phase noise output of -140dBc/Hz at 1kHz. This allowed a refclock frequency of 100MHz (or 400MHz using the

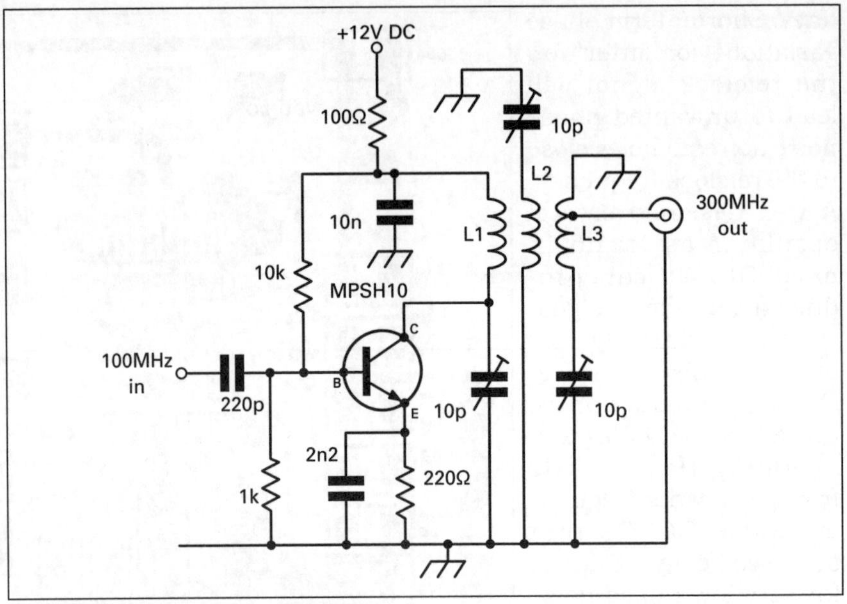

FIGURE 66: Final version of the VHF tripler.

AD9951 PLL-based refclock multiplier). The finished project uses a 300MHz refclock based on the OCXO/tripler circuit described on page 83 (Figure 59). **Figure 66** shows the schematic of the final version of the frequency tripler. The modified circuit has a triple tuned 300MHz output BPF. L1, L2 and L3 are each four turns of 1mm bare copper, wound wide spaced on a 5mm former. The output tap on L3 is 0.8T from the grounded end. Each coil is about 11mm long and spacing between L1/L2 and L2/3 is about 2mm. The 20pF trimmer capacitors from the original circuit have been replaced by 10pF ceramic trimmers. The 300MHz output is only 500mV peak-peak. This has proven to be more than sufficient to drive the AD9951 refclock input.

The initial configuration of the DDS DAC output used a pair of 100Ω resistors as shown in the schematic and overlay. The voltages developed across these resistors was fed via a pair of 10nF capacitors to a simple 4:1 broadband balun as shown in **Figure 67**. The balun is six turns, trifilar wound on an FT37-43 ferrite toroid. I used three lengths of 0.375mm enamelled copper (Maplin YN86T or similar) twisted together.

The frequency tuning word (FTW) of the AD9951 is 32 bits wide. This gives a minimum frequency step of refclock/232. In the case of a 100MHz clock this step is 100000000÷232 = 0.023283064Hz. To calculate the FTW for a given frequency, the formula is FTW = fout/min_step. Typical tuning step values are a small fraction of 1Hz and FTW values are in the hundreds of millions. The PIC microcontroller will handle the tedious business of calculating the FTW value.

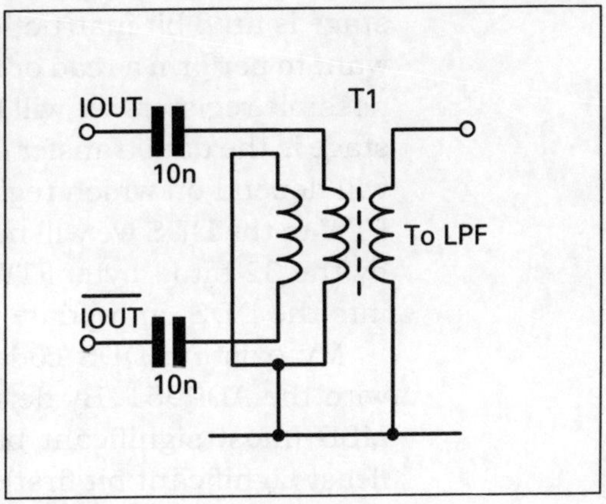

FIGURE 67: Initial DDS output arrangement using capacitive coupling and trifilar wound transformer.

Pinguino PIC Board

The DDS is controlled by the Pinguino PIC board **[24]**. Note that the three DDS pins used for interfacing the PIC to the DDS are named Serial data = SDIO, Serial clock = SCLK, Update = I/O UPDATE. Slightly different names are used for the older AD9851 used in previous projects.

The Pinguino board is based on the well known Arduino board. Both boards can be programmed using the C language or by using a set of simple Arduino-style instructions. In either case, the code for the PIC18F4550 is generated by the SDCC (small device C compiler) **[25]**. If you want to experiment, the PIC board project (Homebrew November 2009) and the AD9851 DDS project (Homebrew March 2010) gave examples of how the Pinguino board can be used to drive an LCD display module, read a numeric keypad, read a rotary shaft encoder and drive the AD9851 DDS.

In order to communicate with the new DDS, I defined three named constants to represent the pins:

```
#define DDS_DATA 13
#define DDS_CLK 14
#define DDS_UD 15
```

Pinguino pin numbers 13, 14 and 15 correspond to PIC18F4550 pin numbers 2, 3 and 4. These constant names are much easier to remember than the obscure Pinguino numbering system. DDS_DATA is connected to the DDS SDIO pin, DDS_CLK is connected to SCLK and DDS_UD is connected to I/O UPDATE.

When I connected the Pinguino board to the DDS, I couldn't resist the urge to apply power to the circuit, just to see what would happen. The Pinguino board was programmed to work with the old DDS project. Not surprisingly, nothing happened. The physical interface is just the same as it was on the previous DDS project, but the AD9951 is programmed in a completely different manner.

I started by calculating the FTW for an output frequency of 10MHz. For a 100MHz clock, fout/min_step = 429496730. This value was assigned to a variable named 'freq'. The SDCC compiler variable type 'unsigned long' is 32 bits wide and has a maximum value of just over 4 billion. That sounds like a lot, but it is just adequate for our purposes. So,

```
unsigned long freq=429496730;
```

Communicating with the DDS is done in two stages. The first

stage is an 8 bit instruction byte. This tells the device whether we want to perform a read or a write operation and which of the several possible registers we will be writing to or reading from. The second stage is the data transfer cycle. The number of data bits transferred will depend on which register we are writing/reading. To send the FTW to the DDS we will need to send the 8 bit instruction, followed by the 32 bit (4 byte) FTW. Once these 40 bits have been clocked into the DDS, an update pulse is applied to the I/O UPDATE pin.

My existing DDS code from previous projects wouldn't work with the AD9951. By default, data is transferred to the new DDS MSB (most significant bit) first. The older DDS takes data LSB (least significant bit first). The first task is to send the instruction byte to the DDS. Going over the data sheet information for the instruction byte bit-by-bit with a pencil shows that the required byte is '00000100'; this is more conveniently expressed as 0x04 in hexadecimal.

The following code is from a C function called dds_inst. The value 0x04 has been assigned to a variable called ctrl.

```
for(count=0;count<8;count++)
{
if(ctrl&0x80)
digitalWrite(DDS_UD,HIGH);
digitalWrite(DDS_UD,LOW);
{digitalWrite(DDS_DATA,HIGH);}
else {digitalWrite(DDS_DATA,LOW);}
digitalWrite(DDS_CLK,HIGH);
digitalWrite(DDS_CLK,LOW);
ctrl=ctrl << 1;
}
```

The first line establishes a loop that will be repeated eight times, once for each bit of the instruction byte. The third line is a conditional test that performs a logical AND of ctrl (the instruction byte) and the binary number '10000000' (0x80 HEX). If the MSB (leftmost bit) of the instruction byte is a 1, the test condition is found to be TRUE and the DDS_DATA line is set high. If the MSB is 0, the test condition is FALSE and the DDS_DATA line is pulled low (line 4). Regardless of the outcome of this test, the next two lines generate a data clock pulse by pulling the DDS_CLK line high and then low. The final line performs a bitwise left shift of the instruction byte using the C left shift operator "<<". This moves the next bit into the MSB position for the next iteration of the loop. After 8 bits have been sent, the loop is terminated.

The next stage of the communication cycle sends the data to the DDS. For writing to the FTW register (FTW0), we will need to send 32 bits. The DDS frequency word is assigned to the variable 'freq'.

```
temp=freq;

for(count=0;count<32;count++)
{
if(temp&0x80000000)
{digitalWrite(DDS_DATA,HIGH);}
else {digitalWrite(DDS_DATA,LOW);}
digitalWrite(DDS_CLK,HIGH);
digitalWrite(DDS_CLK,LOW);
temp=temp << 1;
}
digitalWrite(DDS_UD,HIGH);
digitalWrite(DDS_UD,LOW);
```

This is very similar to the code for sending the instruction byte. The FTW is assigned to a new variable called temp. As with the previous example, a logical AND is used to find the value of the MSB. Note that in this case, we are dealing with the MSB of a 32 bit number. After 32 iterations of the loop, the final two lines send an update pulse to the DDS.

These two blocks of code have been declared as C functions, This makes it trivially easy to re-use them in future code. To show just how easy this is, I can write a new frequency to the DDS by calling the new functions in this sequence:

```
dds_inst(0x04);
dds_data(freq);
```

Testing

My first attempt at testing the DDS was entirely unsuccessful. I had mistakenly assumed I needed to change the CFR1 register bit 12 (Enable SINE Output) to bring the DDS to life. This had the opposite affect and stopped it from working altogether. After much swearing and removing of code line-by-line, I eventually found the cause of the trouble and the offending code was removed.

At the second attempt, a lovely clean sine wave appeared at the DAC current outputs. The output from the transformer in Figure 67 was fed through a 31MHz LPF and then into the 50Ω input of my spectrum analyser. The signal was clean and free of measurable spurs other than the second harmonic at -50dBc. I was pleased to find that the output frequency was correct to within 1Hz. For a second test, the DDS was programmed for the Homebrew Net frequency of 3.727MHz + the IF offset of 10.6985MHz = 14.4255MHz. This time I let Pinguino handle the maths:

```
freq=(unsigned long)
(dial+if_offset)/min_step;
dds_inst(0x04);
// FTW0 register 0x04
dds_data(freq);
// Send 32 bit frequency word
```

The output spectrum over a 100MHz span is shown in **Photo 37**. Second harmonic output is more than 50dB down and all DDS spurs are generally below -70dBc. A 10MHz span centred on 14.42MHz is shown in **Photo 38**. There are no measurable spurs within the critical ±1MHz area. This is in line with data sheet claims of narrow band SFDR (spurious free dynamic range) in excess of 80dB. There are also no measurable spurs at or near the RF and IF frequencies.

The simple RC DAC output circuit seems to work quite well in practice. The low room for improvement. I replaced my simple FT37-43 ferrite toroid. I used four lengths harmonic and spur level suggests that it output network with a broadband transformer of 0.375mm enamelled copper twisted is quite linear, but there may still be some room for improvement. I replaced my simple output network with a broadband transformer that is directly connected to the DAC current outputs and AVDD. The transformer is a centre-tapped 1:1 type. The schematic of the new output network and its LPF are shown in **Figure 68**. This circuit is based on the AD99xx evaluation board schematics from Analog Devices. The value of the two DAC resistors has been reduced from 100Ω to 50Ω. Instead of replacing the 100Ω resistors on the DDS carrier PCB, I left them in place and placed a pair of 100Ω resistors

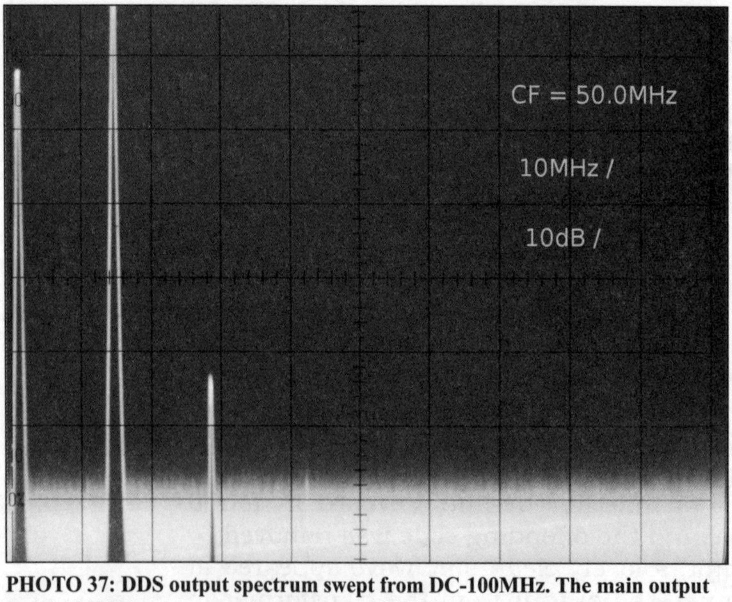

PHOTO 37: DDS output spectrum swept from DC-100MHz. The main output is at 14.25MHz; the second harmonic at 28.5MHz is more than 50dB down.

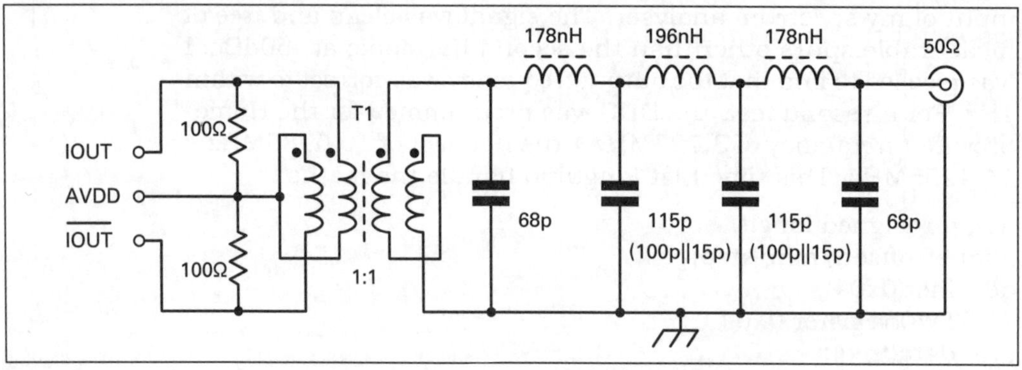

FIGURE 68: Improved output arrangement using quadrifilar transformer and 62MHz low pass filter.

in the output transformer primary. These are effectively in parallel with the on-board 100Ω resistors. The transformer is 6 turns quadrifilar wound on a FT37-43 ferrite toroid. I used four lengths harmonic and spur level suggests that it output network with a broadband transformer of 0.375mm enamelled copper twisted is quite linear, but there may still be some that is directly connected to the DAC current together. As the new arrangement has a direct DC connection to the DAC outputs and AVDD, it is important to check that it is wired correctly and there are no inter-winding shorts.

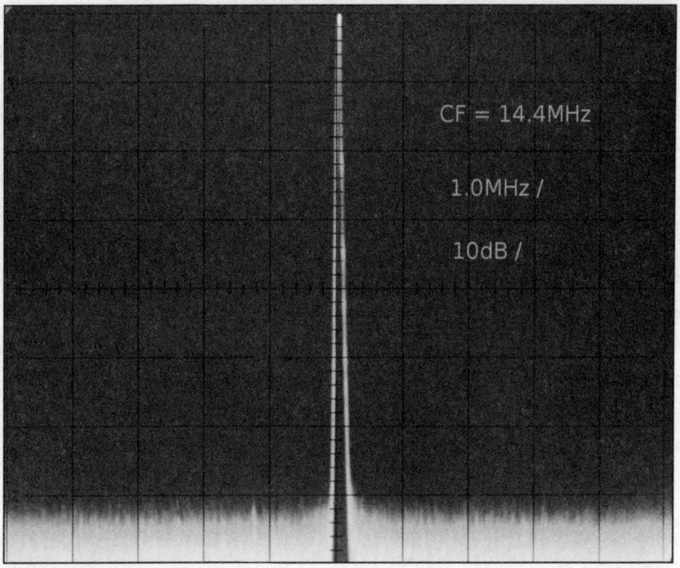

PHOTO 38: A close-in look at the signal, at 1MHz per division, shows that the signal is clean..

The LPF cutoff frequency is 62MHz. L1 and L3 are each 5T of 1mm enamelled the new output copper on a T50-6 toroid. I had to bunch network and its the turns until they occupied about 60% of the core circumference to achieve the required inductance of 178nH. L2 is 6T circuit is based evenly spaced on a T50-6 toroid. **Figure 69** on the AD99xx shows a QUCS simulation of the filter evaluation board response. Measurements of the filter confirm the accuracy of the simulation.

The new output circuit has reduced the 2nd harmonic output to a level between -55 and -58dBc. Spur levels are at least as low as with the previous output network. Side by side tests with the old AD985d1 DDS show that the hew unit has significantly lower spurs. I wrote a simple frequency sweep program for the Pinguino. A YouTube video at **[26]** show how the DDS behaves at output frequencies between 1MHz and 50MH; I used the 300MHz OCXO/tripler as the refclock. The Pinguino source code can be found on the project web page at **[27]**.

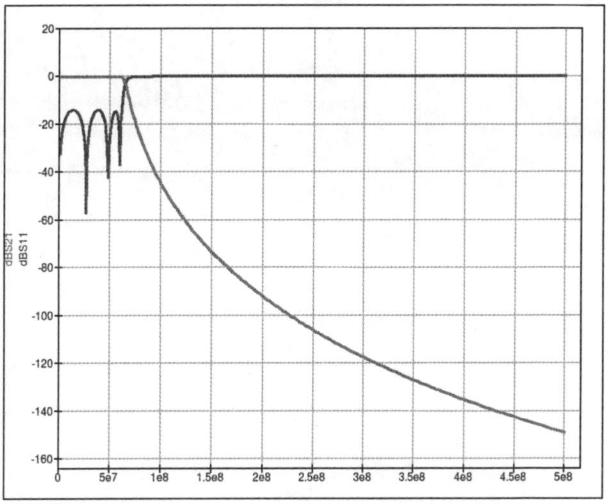

FIGURE 69: Simulated frequency response of the low pass filter. Measurements confirmed the real filter worked as expected.

[23] http://homepage.eircom.net/~ei9gq/art.html
[24] Homebrew November 2009 and http://homepage.eircom.net/~ei9gq/picboard.html
[25] http://sdcc.sourceforge.net/
[26] www.youtube.com/watch?v=flt-zlzK8yk
[27] http://homepage.eircom.net/~ei9gq/ad9951.html

11

Band switching

SINGLE OR MULTI-BAND? Most of the published designs for amateur radio transmitters and transceivers are for relatively simple single band equipment. The additional circuits and switching required for a multiband rig will significantly increase the size and complexity of the equipment. If you look inside a commercially made all-band transceiver, you will find that much of the space inside is taken up by banks of band switched filters and a complex multi-band frequency synthesiser. It is possible to squeeze a multi-band design into a small space, but there are always compromises involved. Small surface mount inductors and capacitors tend to have relatively low Q, high loss, limited power handling and poor linearity. Most designs use conventional toroid core inductors for the output low pass filters and either toroids or slug-tuned inductors for the band pass filters. The low pass filters are usually relay switched because mechanical relays have better linearity and greater power handling capacity than solid-state switches.

Figure 70 shows a simplified block diagram of the transmitter

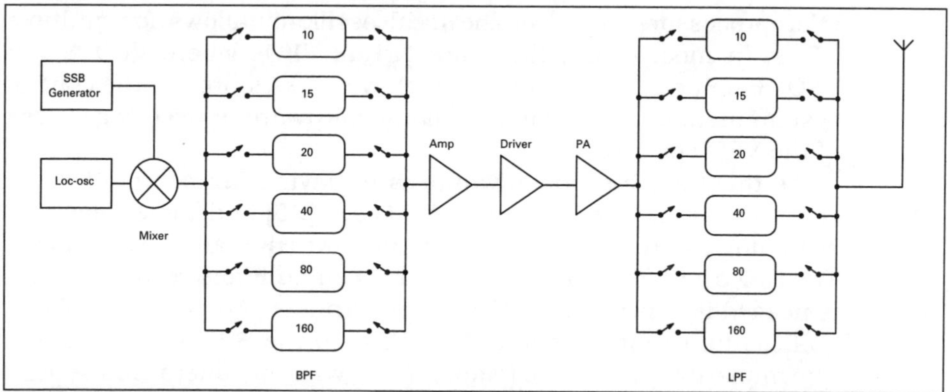

FIGURE 70: Simplified block diagram of a transmitter showing bandpass and lowpass filters and the necessary switching.

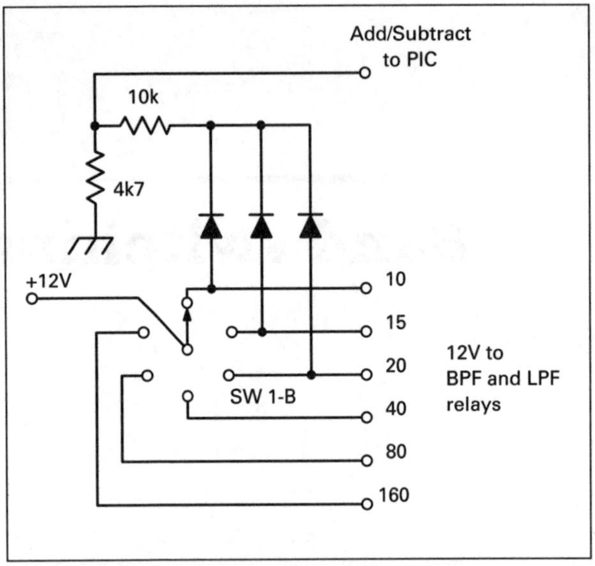

FIGURE 71: Basic band switching solution using a rotary wafer switch to provide DC drive to band relays.

stages of one of my first multi-band SSB/CW transceivers. I used PIN diodes for switching the BPF and a pair of relays for each LPF. A pair of wafer type rotary switches provided band switching. One wafer was used for direct switching of taps on the VFO inductor, the other wafer provided 12VDC for BPF and LPF switching. In this example, there are an equal number of bandpass and lowpass filters. The switch was assembled with one wafer inside the VFO enclosure. **Figure 71** shows the configuration of the DC switching. Note the add/subtract output that tells the PIC based digital frequency display whether the IF offset should be added to (higher bands) or subtracted from (lower bands) the measured VFO frequency. This arrangement worked well in practice and it is still in use in my old (but still current) nine band HF transceiver.

At one time, most amateur transceivers used a rotary switch with a long shaft and several switch wafers for band switching. The switch shaft often ran all the way from the front panel to the valve PA output matching network at the back of the transceiver. This arrangement was popular until the 1980s. The newer generation of transceivers featured microprocessor control and general coverage receivers. As mechanical rotary switches are not easily controlled by a microprocessor, these newer rigs used PIN diodes and/or relays for band switching. Direct microprocessor control of band switching and the local oscillator frequency allows for advanced features like split frequency, cross-band and even remote operation. It also allows you to save any frequency/band combination to memory. Microprocessor control of the local oscillator allows for multiple VFOs. In most cases, these are virtual VFOs where VFO-A and VFO-B are actually just two frequency values stored in the control system memory. The same oscillator hardware is used regardless of the VFO selected.

As there are obvious advantages to having full microprocessor control, I will use the Pinguino PIC board **[28]** and some electronic switching as the band switch for the new HF transceiver project. The BPF and LPF units have already been built and tested. All that is needed is a switched 12VDC supply that is capable of supplying sufficient current to operate the relays in the BPF and LPF units. The miniature 12V relays (Maplin N17AW) in my filter units require DC coil current of 17mA per relay or 68mA for a full set of four (two in the BPF and two in the LPF). To allow for the possibility that

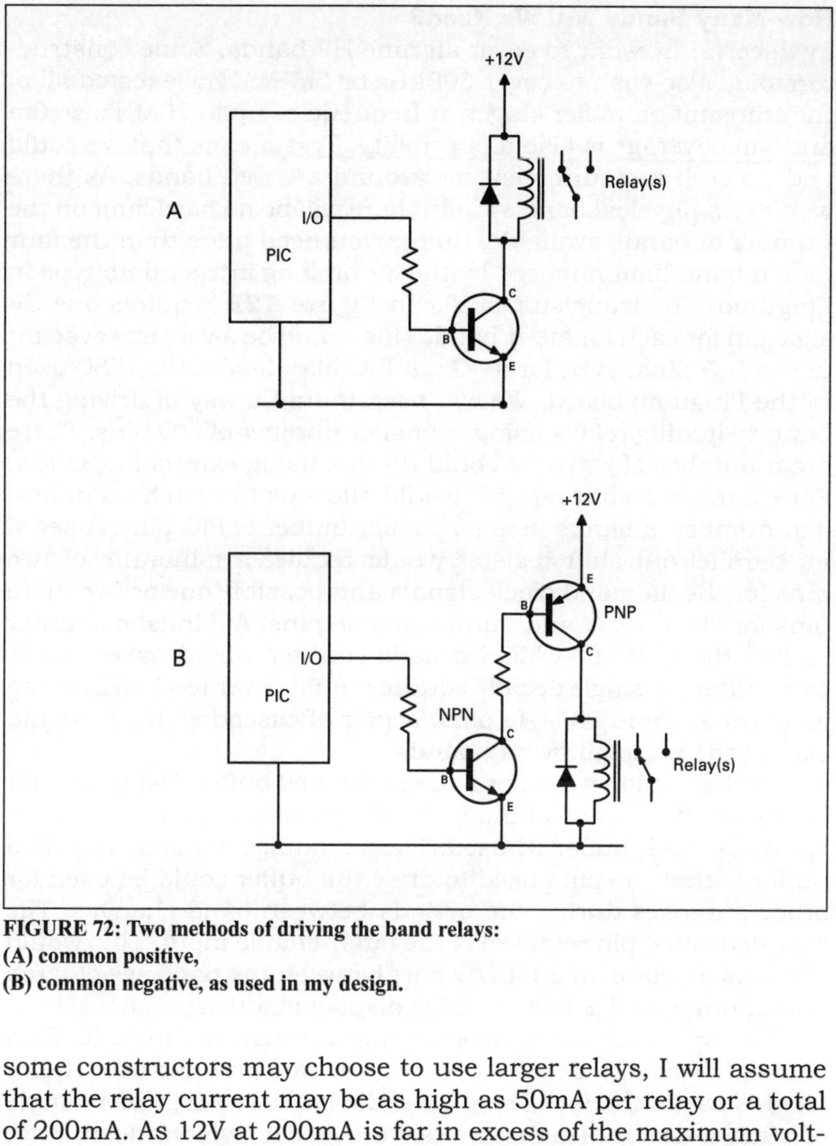

FIGURE 72: Two methods of driving the band relays:
(A) common positive,
(B) common negative, as used in my design.

some constructors may choose to use larger relays, I will assume that the relay current may be as high as 50mA per relay or a total of 200mA. As 12V at 200mA is far in excess of the maximum voltage and current limits of the PIC I/O port pins, it will be necessary to use transistor buffers between the PIC and the relays. **Figure 72A** shows a simple way of driving one or more relays from the 5V available at the digital I/O pins of the PIC. Unfortunately, this arrangement won't work with my filter units because I have wired the relay coils so that the positive supply is switched and the negative supply for all coils is connected directly to ground. Rather than rewiring all of the relays, I have chosen to use the circuit in **Figure 72B**. I used BC547 or BC548 transistors as the NPN device. Just about any general purpose NPN transistor should work equally well. I used the TIP30A power transistor as the PNP device. Any similar PNP power transistor with hFE greater than 20 should be equally suitable.

How Many Bands Will We Need?

I will certainly want to cover all nine HF bands. Some constructors may also wish to cover 500kHz or 5MHz. I have tested all of the transmit amplifier stages at frequencies up to 70MHz, so 6m and 4m coverage is also a possibility. This means that we could end up with a requirement for around a dozen bands. As there won't be a physical band switch, there will be no hard limit on the number of bands available, unless you need more than the four billion band limit imposed by the 32 bit long integer data type in Pinguino. The transistor buffer in **Figure 72B** requires one 5V logic pin for each amateur band. This would be a very extravagant use of I/O pins, even for a 40 pin PIC like the PIC18F4550 used in the Pinguino board. We will need to find a way of driving the band switching relays using a smaller number of I/O pins. There are a number of ways we could do this using external logic ICs. For example: a shift register would allow us to switch an unlimited number of bands using a small number of PIC pins. A serial in, parallel out shift register would require a minimum of two pins for the data and clock signals and possibly one or two more pins for clear/reset and output-enable pins. A Johnson counter IC like the CD4017 CMOS decade counter would have similar capabilities. A single decade counter could cover ten bands using just two or three PIC I/O pins. A pair of cascaded 4017s would extend the range to twenty bands.

Another option would be to use a latched buffer. Using an 8 bit latched buffer on any of the 8 bit PIC I/O ports would cover eight bands for each buffer IC used. The advantage of using a latched buffer is that the pins used to drive the buffer could be used for other purposes during the periods between band changes. The only dedicated pin required is the latch-enable input. This would allow us to share an 8 bit I/O port between the band select latch and another device like the LCD display module.

After spending some time digging through my logic IC data books, I decided that a 4 to 16 decoder would be the best choice for this particular job. The 4514B CMOS IC is a typical example of such a device. This is a 1-of-16 decoder, which means that only one of the 16 outputs is active at any one time. The desired output is selected by 4 bit binary inputs with a value of 0 to 15 (or 0000 to 1111 in binary). I used a Philips HEF4514BT. The 4514B is a latched device, so the inputs only need to hold the correct value for a few hundred nanoseconds before, during and after the latch-enable pulse. This means that the four addressing pins are free for other purposes between band changes. The only PIC pin dedicated to band changes is the I/O pin used to drive the latch-enable input. This allows us to cover up to sixteen bands with just a single PIC pin.

The PIC PORTA,5 pin is used to control the latch-enable. This is pin 7 of the PIC18F4550 or digital I/O pin 17 in the Pinguino numbering system. Its value is defined in the Pinguino source code as:

#define band_latch 17

This allows us to address this pin as "band_latch" in our Pinguino code. For example: digitalWrite(band_latch,HIGH); We must supply a 4 bit binary number representing the band number to the 4514B input pins. The band numbers have also been pre-defined in the Pinguino source code as 0=band_160, 1=band_80, 2=band_40 etc. These values are easily changed to suit your individual requirements. Once again, we have several options available. When I wrote the original LCD module code for the Pinguino, I decided to use an 8 bit interface to the 2x16 LCD module instead of the 4 bit interface used in earlier projects like the PIC frequency counter (October 2006). The LCD module data inputs are driven by the 8 bit PORTB (PIC pins 33-40). Driving the LCD module in 4 bit mode would leave four PIC pins free for driving the 4-16 band switch decoder. The cost of this change would be a doubling of the amount of time it takes to write each character to the LCD module. As the LCD module shows the dial frequency of the transceiver and this information must be updated every time the VFO frequency is changed, I decided that I would not take this approach. Sluggish VFO tuning is just too high a price to pay.

I eventually decided to drive the band decoder using the lower four bits of PORTB. These pins are shared with the LCD module. As the 4514B decoder IC and the Hitachi (or compatible) LCD module both have latched inputs, their inputs can bemultiplexed, provided that only one device is written to at any time. **Figure 73** shows how the band decoder is configured.

Writing to the decoder is a simple matter of placing the desired band number on PORTB pins 0, 1, 2, 3 and generating an enable

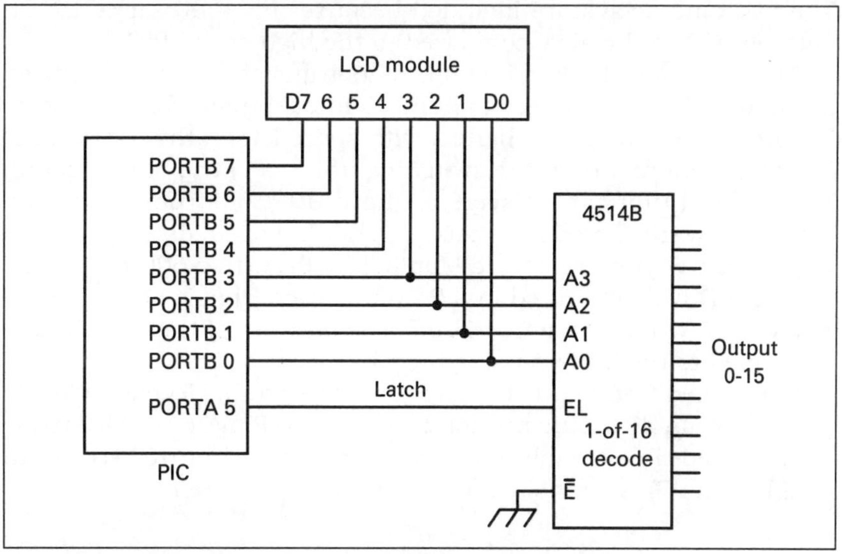

FIGURE 73: How the band decoder and LCD share PORTB pins 0-3.

```
/* Band switching */
        void change_band(unsigned char band)
        {
        PORTB = band;                        // 1-of-16 decoder inputs PORTB 1-3
        digitalWrite(band_latch,HIGH);       // Latch enable pulse
        delayMicroseconds(10);               // Allow time for slow CMOS logic
        digitalWrite(band_latch,LOW);
        delayMicroseconds(10);
        }
```

FIGURE 74: Simple C function to write to the band decoder.

pulse on PORTA,5 (band_latch). The code to do this has been written into the very simple C function in **Figure 74**. Changing band is as simple as calling this function in the following manner:

change_band(band_20);

This will select the appropriate decoder output for the 20m band. The first line puts the band number on PORTB

PORTB=band;
The next line generates the latch-enable pulse.

digitalWrite(band_latch,HIGH);

This is followed by a 10μs delay before the latch-enable line goes low.

delayMicroseconds(10);
digitalWrite(band_latch,LOW);
delayMicroseconds(10);

The two time delays are included because the 4000 series CMOS decoder IC is quite slow compared to the high speed PIC.

The band decoder and just one of its transistor buffers is shown in **Figure 75**. You will need to build identical buffers for each band. My prototype covers nine bands, with space left on the board for at least three more. I built the switching unit on a strip of Veroboard. The finished unit can be seen in **Photo 39**. Construction is fairly straightforward. The 560Ω resistors should have a dissipation rating of 500mW or more. I used Maplin M560R with a 600mW rating. The 4514B IC is mounted in a 24 pin socket. This is actually a 32 pin socket with a few pins cut off. The outputs of the decoder are connected to the buffers using the white/orange insulated hookup wire (stripped from 4 way telephone cable) and the decoder address and latch enable inputs are connected to the Pinguino board using the coloured ribbon cable at the left of the photo. The 4514B output enable (pin 23) is connected to ground using a wire link.

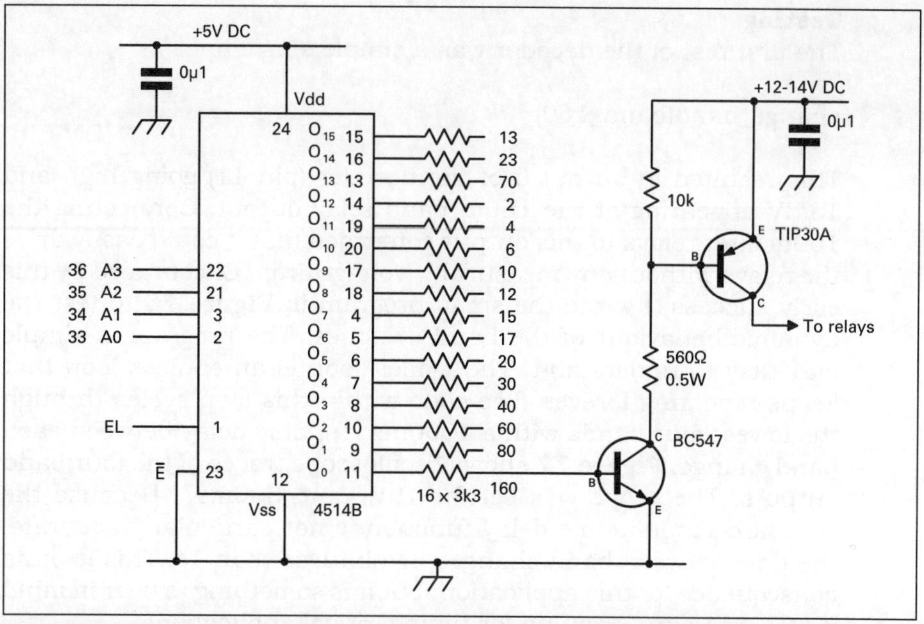

FIGURE 75: Circuit diagram of the band decoder. Note that only one output stage is shown; there can be up to 16 sets of BC547, TIP30A and associated resistors, depending on how many bands are to be covered.

PHOTO 39: Prototype band decode and relay driver board.

Testing

The first test of the decoder was a simple single line:

change_band(band_160);

This resulted in output 0 of the decoder (pin 11) going high and 13.8V appearing at the 160m band relay output. Connecting the 160m filter relays to this output confirmed that it could easily drive the relays without any measurable voltage drop. Encouraged by this early success, I wrote the small program in **Figure 76** to test the dynamic behaviour of the band switches. The program is simple and easy to understand. The while() loop is an endless loop that keeps repeating forever. The code within this loop cycles through the lowest four bands with a 1000ms (1s) time delay between each band change. **Figure 77** shows oscilloscope traces of the four band outputs. The scope was set for 10V/ and 500ms/. Because the standard Pinguino ms delay function is not particularly accurate, the time between band changes is a bit less than 1s. This is of no consequence for this application, but it is something to bear in mind if you are using Pinguino for timingcritical applications.

```
while(1){
  change_band(band_160);
  delay(1000);
  change_band(band_80);
  delay(1000);
  change_band(band_40);
  delay(1000);
  change_band(band_30);
  delay(1000);
}
```

FIGURE 76: Test code to check the dynamic behaviour of the band switches.

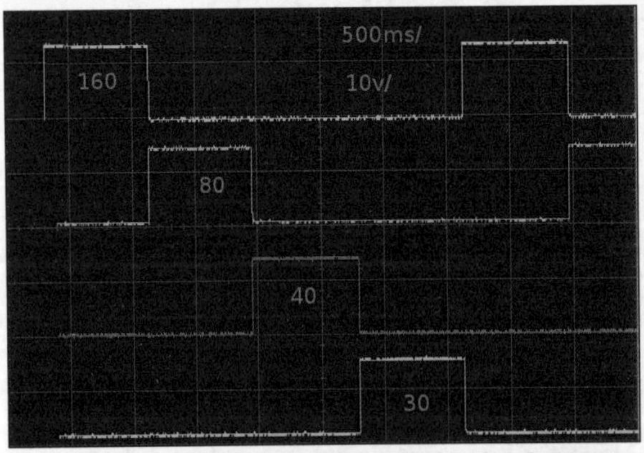

FIGURE 77: Oscillogram monitoring the four band outputs switched by the code in Figure 76.

The code in **Figure 78** was used to test all nine band outputs. As well as testing the switching for every band, this program writes several characters to the LCD module between band changes. This proves that there is no unexpected interference between the band changing and LCD writing functions. You can change the for() loop conditional test parameter ";n<9;" to some other value if you are using more or less than nine bands. The code comments shown in blue give a brief explanation of each line of code.

The value of "band_160" is defined as zero earlier in the source code. This value is assigned to n for the first iteration of the for() loop. The "n++" statement increments the value of n for each of the following iterations. The LCD displays

"Band #0", "Band #1", "Band #2"

etc, returning the cursor to the home position at the left side of the display after each band change. The

```
    lcdinit();                   // Initialise LCD module

    while(1)                     // Endless loop
          {

          for(n=band_160;n<9;n++) // Step through nine bands
           {
           change_band(n);             // Change to next band
           home();                     // Cursor to home position
           display_string("Band #"); // Display text string
           display(n+0x30);            // Convert to ASCII and display
           delay(1000);                // 1000ms delay
           }
          }
```

FIGURE 78: Test code to exercise all 9 band outputs and the LCD.

test showed that output number 6 (band_15) was not working. This was due to a tiny whisker of solder between the base and emitter pins of the BC547 transistor for this band. Once the solder bridge was cleared, all band switches and the LCD display routine worked perfectly. The LCD code is explained in **[29]**. Once I was satisfied that all was working properly, I increased the delay loop to 10s "delay(10000)" to reduce wear on the relays and connected the BPF and LPF units to the new band switch. The switching circuit was tested in this configuration for several hours. Unlike my old six band rig, this rig does not have an equal number of bandpass and lowpass filters. Some of the lowpass filters are used for two bands. **Figure 79** shows how the 10m filter is also used for 12m. The same LPF is also used for 40/30m and 17/15m.

We now have working Pinguino code to drive the AD9951 DDS,

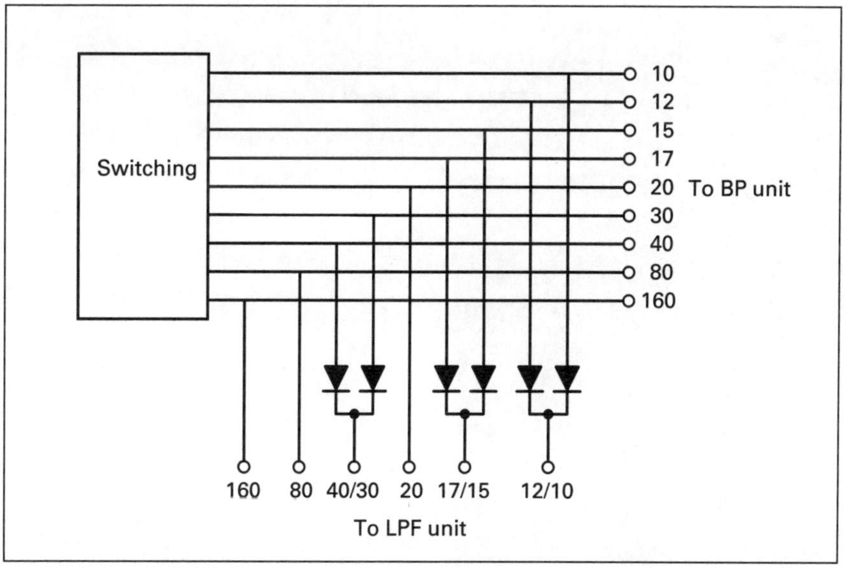

FIGURE 79: How I decoded the bandpass filter drives to feed the smaller number of low pass units.

read a rotary encoder, scan a numeric keypad, drive the 2x16 LCD display and operate the band switching relays. All of the key modules have been built and tested. There are just a few details to tidy up. One obvious omission is the lack of a method of generating CW.

[28] Homebrew November 2009, March 2010
[29] http://homepage.eircom.net/~ei9gq/dds.html

12

Broadband driver amplifiers and generating CW

MORE POWER - The 10W MOSFET PA (chapter 5) has an almost flat frequency response from LF to 100MHz. The cost of this great bandwidth is significantly reduced gain. A combination of negative feedback and resistive attenuation in the input circuit has reduced the gain to a mere 10dB. If this PA is to be used in my HF transceiver, I will need to build a driver amplifier with a gain of approximately 10dB and maximum output of about 1W. My proposed layout consists of a low-level amplifier (page XX) at an output of 100mW, a new driver amplifier with a gain of ~10dB followed by the 10W MOSFET PA. This arrangement would produce more than 10W on all bands. As with the PA section, two alternative designs will be built: the first based on a bipolar transistor, the second on a MOSFET.

Key Components
Driver and PA transistors designed for use in 27MHz CB equipment are a very popular choice with both homebrewers and commercial equipment manufacturers. Many designs are based on the 2SC2166 **[30]** or 2SC1306 driver transistor. These devices have proven to be good performers on all HF bands. They can also be used (with reduced performance) on 6m.

Most of the commonly available RF driver transistors have a maximum dissipation rating of several watts. As the required output level is only 1W, I can easily afford the luxury of running the driver amplifier in Class A for improved linearity. As I have a limited stock of RF driver transistors, I would like to ensure that I don't let the smoke out of them. Active bias circuits will be used in both of the amplifier projects. Tight control of DC bias conditions will ensure that transistor dissipation is kept within predefined limits at all times.

Active Bias

A simplified diagram of an RF amplifier with active bias is shown in **Figure 80**. A resistor in the collector circuit is used as a means of measuring collector current. The voltage drop across this resistor is compared with a reference voltage. In this case, the reference is a pair of forward biased silicon diodes. As each diode has a forward voltage drop of around 0.7V, the voltage across the pair is approximately 1.4V. The reference voltage is almost completely independent of transistor current and transistor temperature. The circuit will try to maintain transistor current at a level that produces a voltage drop across the current-sense resistor that is identical to the reference voltage. If we want to build a Class A amplifier with a continuous current of 1A, we would use a resistor value of V_ref/I or 1.4/1 = 1.4Ω.

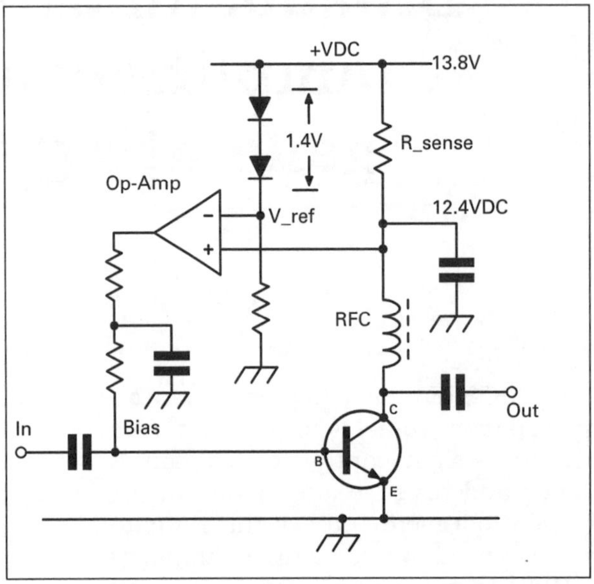

FIGURE 80: Basic amplifier with simplified active bias circuit.

Opamp circuits can be fussy and unpredictable, especially when they are

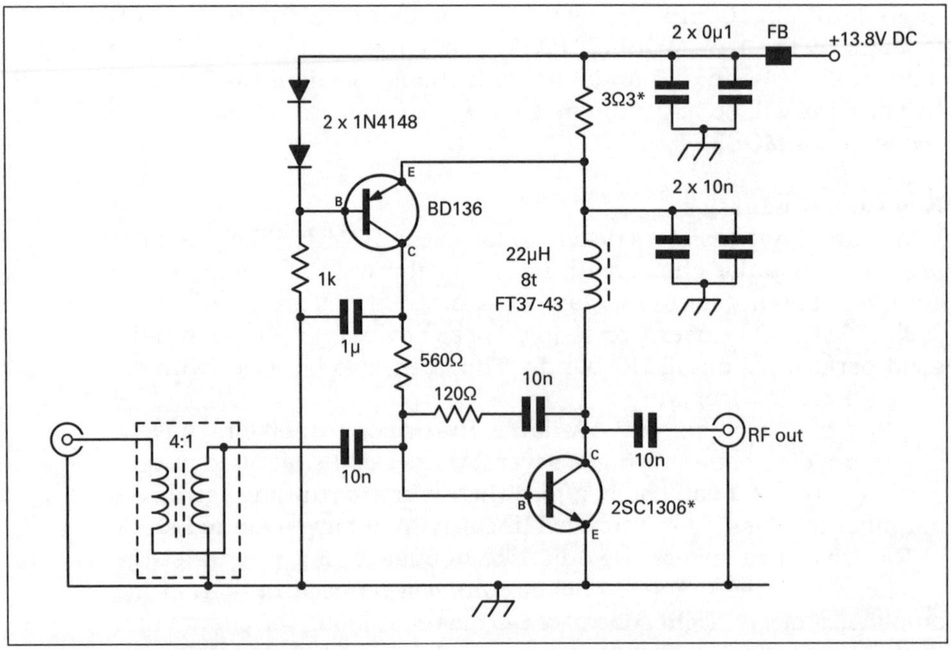

FIGURE 81: Bipolar 1W driver amplifier.

mounted in the same enclosure as an RF power amplifier. Opamp 'latch-up' or any other unpredictable behaviour could lead to excessive bias voltage to the PA transistor. It would be a terrible irony if our new driver amp was blown up by its own protective circuitry. I have decided to use a much simpler circuit based on a PNP transistor. **Figure 81** shows the final version of the Class A driver amp, described later. This amplifier also uses a pair of diodes as the active bias voltage reference. The inverting input is the transistor base, the non-inverting input is the emitter. Open-loop gain is less than the gain of an opamp IC, but it is still more than adequate for a simple bias regulator. When calculating the value of the current-sense resistor, we must allow for the voltage drop across the transistor base-emitter junction. This is typically around 0.6 to 0.75V. **Figure 82** shows measured voltages from one of my prototype amplifiers. The reference voltage is 1.45V (2x0.725). Forward voltage drop of the B-E junction of the PNP transistor is 0.68V. The current sense resistor is 3.333Ω, which is made from three 10Ω resistors in parallel. The measured voltage across this resistor is 0.77V. Note that the sum of these last two voltages is exactly the same as the reference voltage.

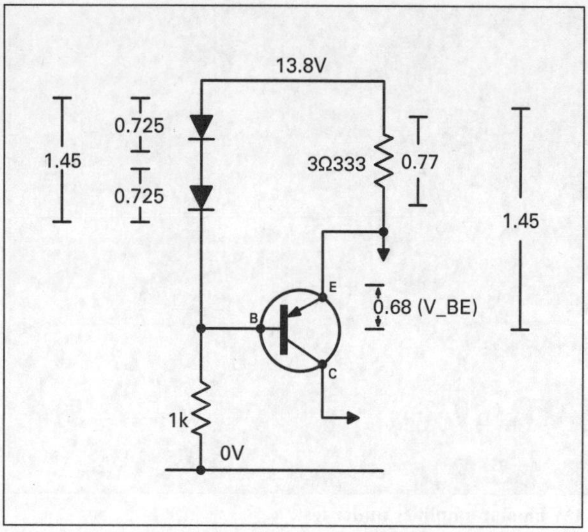

FIGURE 82: Measured voltages on a prototype amplifier.

A 1W Bipolar Linear Amplifier

The first of our four projects in this Chapter is a driver amplifier based on a 2SC1306 bipolar transistor. In my prototype, the transistor is mounted on a heatsink that is much larger than necessary. As the heatsink tab is internally connected to the collector, I had to use an insulated mounting kit. At the lowest frequency of operation, the collector choke should have an inductive reactance several times the 50Ω output load impedance – at least 200Ω. A 22.4μH inductor made from 8 turns of insulated wire on an FT37-43 ferrite toroid meets this requirement. AL=350. $AL*N^2 = 350*8^2 = 22{,}400$nH. 2πfL = 2*π*1.8*22.4 = 253Ω, which is about 5 times the load impedance at 1.8MHz – and much greater at higher frequencies.

The first test circuit had no 4:1 input transformer and no negative feedback from collector to base. Measured gain was well over 30dB at 1.8MHz, falling to 10dB at 40MHz. Adding a simple feedback network consisting of a series connected resistor and capacitor from

PHOTO 40: 1W bipolar amplifier under test.

collector to base improved gain flatness considerably. Gain with a 120Ω feedback resistor was about 12dB across the HF bands, falling to 10dB at 2MHz and 40MHz. Input matching to 50Ω was very poor. Adding a simple 4:1 transformer to the input circuit improved the input return loss (RL) to about 6dB (SWR=3:1) or better. The 4:1 transformer is 10T of bifilar-wound 0.375mm enamelled copper wire on a FT37-43 ferrite toroid. This modification also improved gain slightly to 14-15dB from 1.5-30MHz, falling to 10dB at 40MHz. This would probably make a good 1W driver amplifier for a 160-10m rig. As the gain is slightly greater than required, the input match could be improved by using a simple input attenuator of 4dB or so. This would improve input RL to about 14dB (SWR=1.5:1). However, the lack of gain at 50 and 70MHz makes it less than ideal for our project. Figure 81 shows the amplifier schematic; **Photo 40** shows the amplifier under test.

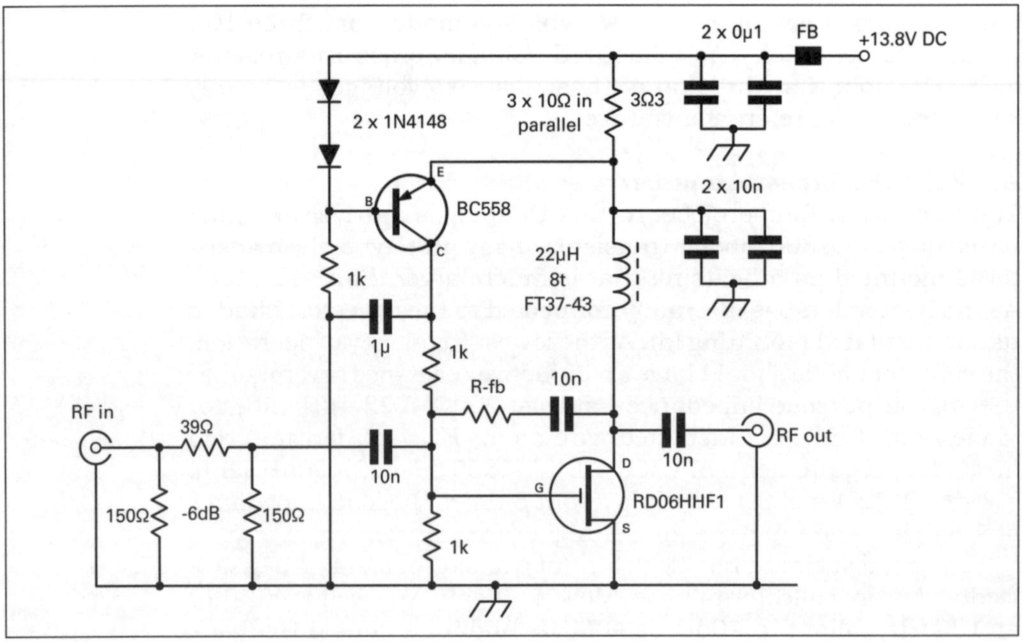

FIGURE 83: 1W MOSFET amplifier.

A 1W MOSFET Linear Amplifier

For the next attempt at building a 1W driver amplifier, I used a Mitsubishi RD06HHF1 MOSFET **[31]**. The amplifier schematic is in **Figure 83**. As with the previous amplifier, the prototype was built on an unnecessarily large heatsink. One of the many nice things about the RDxx series of HF MOSFETS is that the heatsink tab is connected to the FET source terminal instead of the drain. In the usual case of a commonsource amplifier where the negative supply is tied to the amplifier chassis and heatsink, there is no need to insulate the FET from the heatsink. The heatsink was drilled and tapped so that an M3 screw could be used to mount the FET. I used a very thin film of heatsink compound between the FET and heatsink. The choke in the drain circuit is identical to the one used in the earlier bipolar design.

For the first tests, I used a 220Ω resistor in the negative feedback network (R_fb). This amplifier also used active bias control. I didn't have a 3.3Ω resistor available, so I used three 10Ω resistors connected in parallel as the current-sense resistor. A Class A amplifier should be biased for DC input of significantly more than twice the RF output power. I chose to bias the amplifier for a standing current of around 250mA, for a DC input of just over 3W. As with the earlier bipolar amplifier, I found the voltage drop across the 3.3Ω resistor was 0.77V, indicating FET drain current of just over 230mA for a DC input of 3W.

My first prototype didn't have an input attenuator. Input coupling to the FET gate was via a 10nF capacitor. Testing with a signal generator and spectrum analyser showed that the amplifier has a remarkably wide bandwidth. Gain is 14dB and flat within 1dB from 2MHz to well over 100MHz. The -3dB BW (gain=11dB) is 1.2MHz to 350MHz. Gain falls to 10dB at 0.9MHz and 408MHz. This is far better than expected. **Photo 41** shows the amplifier gain from LF to 100MHz. This is just about ideal for a HF+6m+4m transceiver. Gain actually increases slightly to about 15dB in the 200- 300MHz region before it gradually tapers off above 400MHz. Input return

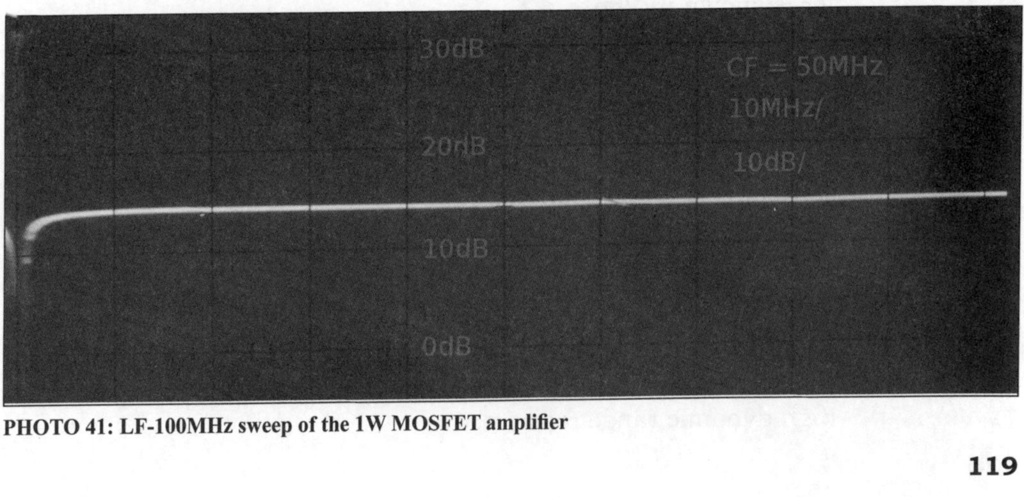

PHOTO 41: LF-100MHz sweep of the 1W MOSFET amplifier

PHOTO 42: Prototype 1W MOSFET amplifier.

loss for the first prototype was about 6dB (SWR=3:1). Increasing the value of the feedback resistor to 700-1000Ω improved input return loss to better than 20dB (SWR=1.2:1), but gain was increased to well over 20dB and gain flatness was also compromised. After some experimenting, I found that a feedback resistor of 560Ω gave a good compromise between gain, bandwidth and input return loss. The final version of the amplifier has a 6dB PI attenuator at the amplifier input. This reduces gain to the original level of 14dB. Input return loss is well over 20dB (SWR=1.2:1) from LF to 84MHz. The amplifier output return loss is 10dB (SWR=2:1). In this configuration, the -1dB points (gain=13dB) are at 2MHz and 80MHz. The -3dB points (gain=11dB) are at 1.2MHz and 150MHz. 3rd order intermodulation products were measured at 40dB below PEP for an output power of 1.3W and 50dB below PEP at 500mW output. As is typical for MOSFET amplifiers, higher order IMD products (5th, 7th order etc) were below the analyser noise floor. The prototype MOSFET amplifier is shown in **Photo 42**.

As the prototype driver amp had shown such good performance over such a remarkably wide bandwidth. I decided to build a second amplifier, just to see if the design was reproducible. I'm happy to report that the second amplifier offers identical performance to the original. I also tested some of the prototype MOSFET driver amplifiers as receive preamplifiers on the 10m band. I was surprised to find that the noise performance compared favourably with my existing HF preamp, which has a noise figure of around 6-7dB. Many American stations were received during these tests. This is remarkable when you consider the high standing current of 230mA and an estimated 3rd order output intercept point of +45-50dBm. It would seem that the RD06HHF1 is an ideal device for use in a high dynamic range receiver.

Generating CW

Most of the key blocks for the new HF transceiver are now built and tested. A key missing item (pardon the pun) is a method of transmitting CW.

There are several ways of generating a CW signal. The traditional method of generating a carrier wave is to use an oscillator. This can be a simple crystal oscillator, a free running VFO or a frequency synthesiser like our recent DDS project (Chapters 9 & 10). The carrier can be keyed directly by switching the oscillator on and off, or indirectly by keying a buffer amplifier stage. It is also possible to generate a carrier wave by injecting an audio tone into the modulator of a single-sideband transmitter. If the modulating tone is a perfect sine wave, the resulting RF signal will be indistinguishable from a carrier wave generated by any other means. To achieve good results, the SSB transmitter must have excellent linearity, opposite sideband suppression and carrier suppression.

The next project is a simple CW generator. The carrier is generated by a simple crystal oscillator running at the transmitter's IF frequency. The oscillator is followed by a two stage buffer amplifier that is keyed by a Morse key. I use a simple PICbased electronic keyer and iambic paddle. The keying circuit also includes semi-break-in switching that is used to control the transceiver Rx/Tx relays. The schematic is shown in **Figure 84**.

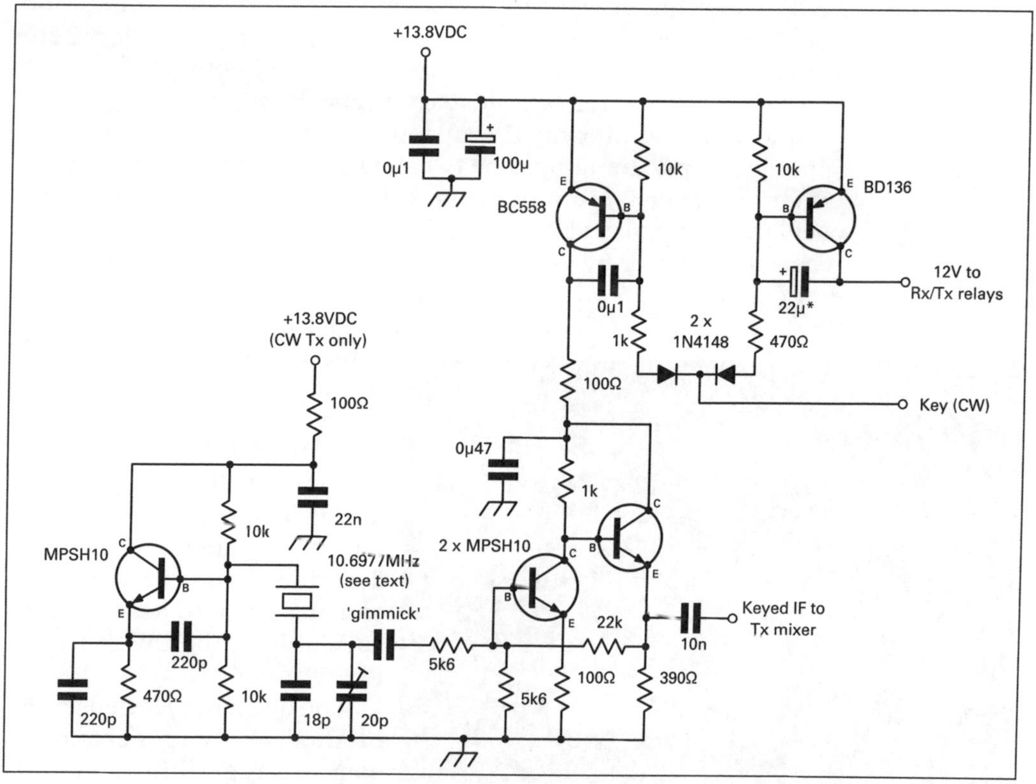

FIGURE 84: Simple IF CW generator with Tx/Rx keying.

I used three MPSH10 NPN transistors for the oscillator and buffer amplifier. A pair of PNP transistors are used for keying and semibreak- in switching. The oscillator is a simple common-collector Colpitts type. The crystal frequency is chosen to match the IF carrier frequency of the HF transceiver. For a receive IF carrier frequency of 10.697MHz, which is just outside the lower edge of the SSB filter passband, I use a crystal running at 10.6977MHz (because I like to listen to CW with a beat-note of around 700Hz). The oscillator output is taken via the crystal. This reduces noise and harmonics at the oscillator output. Coupling to the buffer amplifier is via a very small value 'gimmick' capacitor. This is just two short lengths of insulated wire twisted together. I found just a couple of millimetres of twisted pair (1pF approx) provided enough drive to produce several hundred millivolts at the output of the buffer amp.

A BC558 transistor is used as the keying switch. The 100nF capacitor between the collector and base of this transistor and the relatively large value of the buffer amplifier decoupling capacitor gives slow switching speed. This delay of a few milliseconds on the keying rise and fall times reduces key clicks on the transmitted signal. The semibreak- in circuit uses a BD136 power transistor. A heatsink is not required for this component. The much larger value of capacitor used in this circuit gives a delay of just under 1s after key-up before the transceiver switches back to receive. 22µF is just about right for my style and speed of CW operation. The delay is easily changed by using a smaller value for faster switching or a larger value for even slower switching.

The CW generator was tested by shorting the keying line to ground and monitoring the output using an oscilloscope. The 'gimmick' capacitor should be tweaked so that the output from the buffer amplifier is no more than a few hundred mV. The crystal oscillator frequency should be adjusted so that it is offset from the receiver IF carrier by 700Hz to 1kHz depending on your own preference. **Figure 85** shows the keying waveform (top) and IF output (bottom) from the CW generator. The transmitter is keyed with the letter V. The resulting signals sounds clean and free from key clicks on an SSB receiver tuned to 10.697MHz USB. The prototype CW generator does not have a sidetone oscillator. The audio sine-wave oscillator described in our final projects can also be used as a sidetone oscillator for the CW generator in Figure 84.

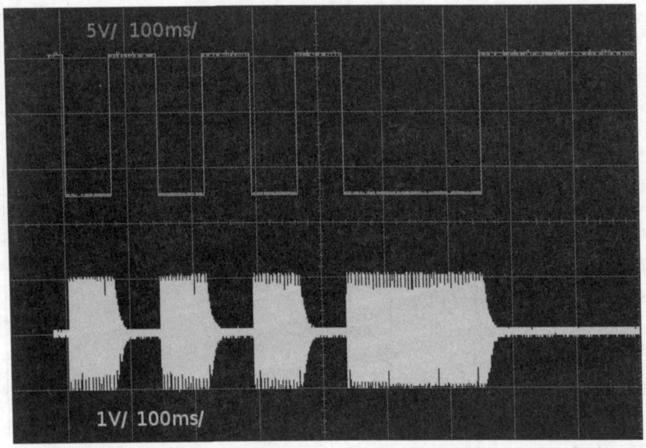

FIGURE 85: Keying waveforms from the simple CW generator.

The oscillator/keyed-buffer method of CW generation is easy to apply and generally well behaved. Providing that the keying rise times are well controlled, key clicks will be kept to a minimum. The sine wave output will be free of any spurious signal except for harmonics. If necessary, these are easily removed by a LPF or BPF. For the ultimate in spectral purity, the CW output can be passed through a narrow bandwidth (1kHz or less) crystal filter. This would remove all harmonics and provide very effective suppression of key clicks at frequencies beyond ±500Hz. If your rig uses a home made crystal ladder filter, it would be quite easy to make a dedicated filter for use in the CW transmitter stages. A simple two or three pole filter should provide excellent results.

My preferred method of CW generation is to inject a keyed audio sine wave into the mic-amplifier/modulator stages of an SSB transmitter. It is more difficult to achieve good results using this method, but it does offer a few advantages over the method used in our previous project. Audio injection is equally effective in upper and lower sideband modes. The Rx/Tx offset is always correct in either mode. This is very useful if you use cross-mode SSB/CW operation on the LF bands (LSB) and the HF bands (USB). The sine-wave injection oscillator can also be used as the sidetone oscillator. This guarantees correct frequency netting. If the received station is at the same pitch as the sideton oscillator, you know that you will be transmitting on the correct frequency.

I decided to use a simple opamp based phase shift oscillator for CW generation. The oscillator is shown at the left side of **Figure 86.** The circuit is based on a simple three stage RC phase shift network (PSN). Each RC pair gives a phase shift of 60° at the oscillator output frequency. The total phase shift of 180° results in positive feedback when applied to the inverting input of the opamp. The output is taken from the final stage of the phase shift network instead of the more usual practice of taking the signal from the output pin of the opamp.

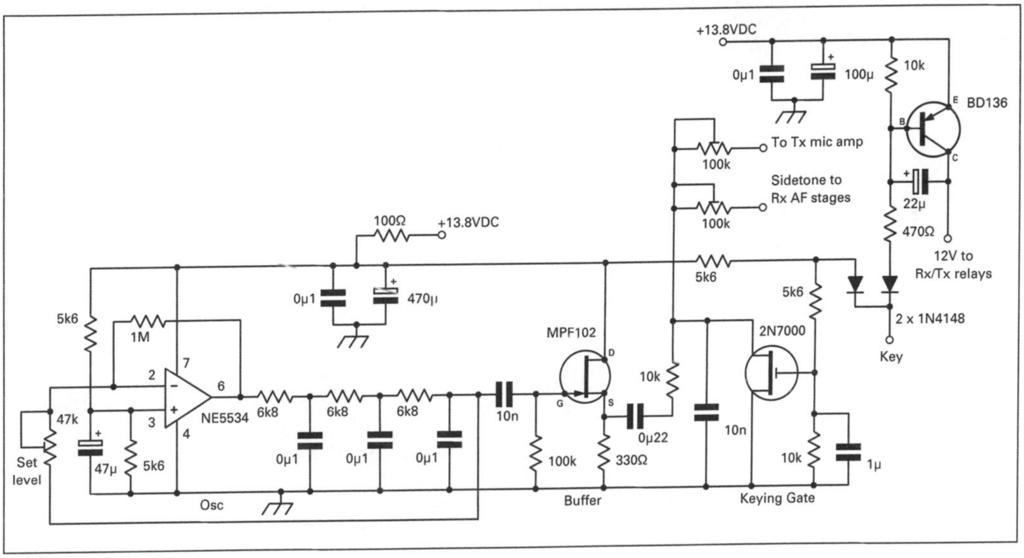

FIGURE 86: Audio CW generator with Tx/Rx keying.

The PSN provides low-pass filtering of the output signal, which gives a useful reduction in harmonic output. The 47k pot is used to adjust the gain so that oscillation is just maintained and the oscillator starts up reliably when power is applied. With careful adjustment, it is quite easy to achieve harmonic suppression of 50-60dB. The oscillator output frequency is around 700Hz. My prototype runs at 680Hz. The oscillator output is fed to a FET source-follower that acts as a buffer amplifier with a high input impedance. The keying gate is a 2N7000 MOSFET that is used as a switch. When the MOSFET is forward biased, the AF signal is shunted to ground. The 2N7000 has a specified ON resistance (RDS) of around 2Ω. The 10k series resistor and MOSFET shunt switch gives signal attenuation of up to 20log(2/10000) = 74dB. As this is greater than the carrier and opposite sideband suppression of a typical SSB transmitter, the signal can be considered as being completely non-existent when the Morse key is up. The semi-break-in circuit has been copied from the previous project. A sample of the output signal can be found at http://homepage.eircom.net/~ei9gq/ dots.mp3.

[30] 2SC2166 data sheet, Mitsubishi, 2007.
[31] RD06HHF1 data sheet, Mitsubishi, 2011.

13

Chassis bashing

As ALL OF THE main modules of the HF transceiver are built and tested, it is time to start looking for a suitable enclosure for the new rig. There are a number of options available. I could build a new enclosure from scratch or the second option is to use an existing enclosure that has been recovered from surplus equipment.

Old industrial, medical or communications equipment is a good source of high quality enclosures. Power supplies and battery chargers are another good source. The third option is to buy a new enclosure. These are available in various sizes. The most popular type is a simple box where the bottom half is made from aluminium and the top cover is made from vinyl covered steel.

Buying a new enclosure is often a good option when you are building a smaller singleband rig. Most if not all of the readily available enclosures are a bit too small for larger projects like a multi-band transceiver.

I eventually decided to use a 19" rack mount enclosure that was recovered from some scrap broadcast equipment. The scrap unit contained some very useful components including a 24V power supply and some nice RF components. These were removed carefully and added to the EI9GQ junkbox. **Photo 43** shows the stripped-down enclosure.

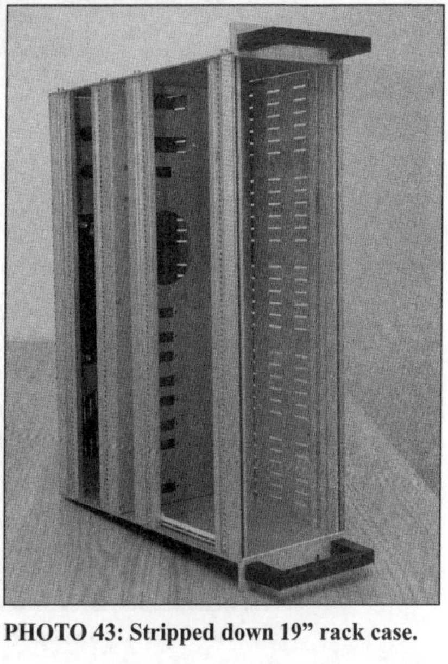

PHOTO 43: Stripped down 19" rack case.

Unless you are very fortunate, the front and rear panels of a surplus enclosure will already have several holes in them. The front panel of a HF transceiver usually has a fairly complicated layout. You will need rectangular holes for the LCD display module, signal meter and keypad. Other holes will be required for the various knobs, switches, buttons and the microphone socket. The back panel will have a socket or wires for the DC power supply, at least one coaxial connector for the aerial and possibly other I/O connections for driving a transverter or other external equipment. It is very unlikely that any existing holes will be suitable for use in your project. In most instances, it will be necessary to make new front and rear panels. These can be screwed directly onto the original panels or, in the case of rack mount enclosures, the original panel can simply be removed and replaced by a new panel.

Cutting rectangular holes is one of my pet hates. Getting the edges straight and the corners square can be difficult with an aluminium panel and almost impossible with steel. A good sheet metal nibbling tool will make your job a lot easier. I usually use sheet aluminium for home made panels and enclosures. In the past, I have used a hacksaw for cutting the aluminium. I always found this method slow, tedious and not very accurate. I have an electric jigsaw, but this usually gives even worse results than a hand saw. In recent years, I have found that the score and bend method gives much better results. To make straight and accurate 90° bends, simply cut a deep score in the aluminium sheet using a sharp tool like a carpet knife. The sheet is then bent along the line of the score using a sharp edge like a metal workbench. With a little practice, you will find that it is easy to make perfectly

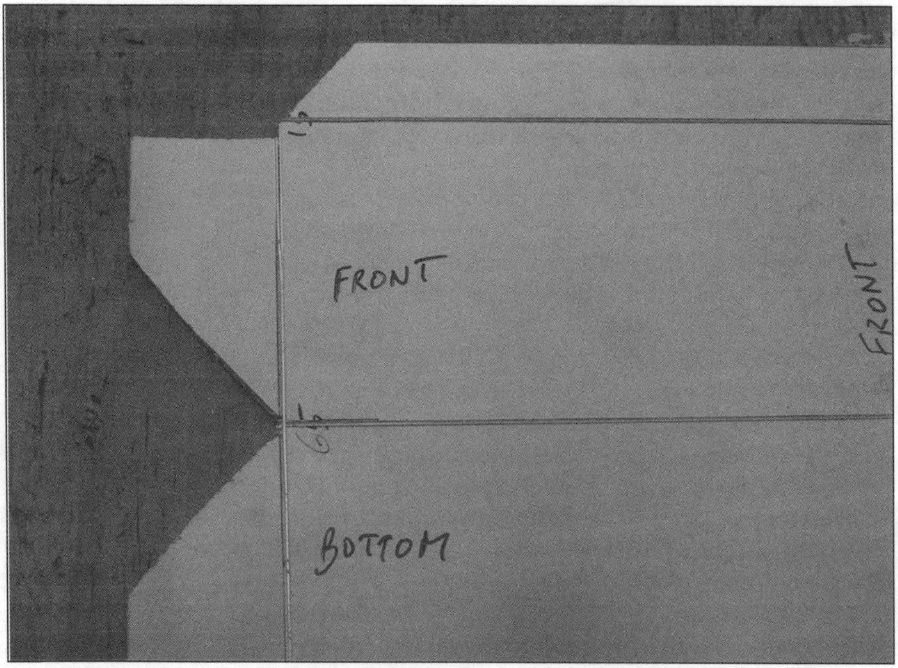

PHOTO 44: Metal panel marked out and scored prior to bending.

straight bends. If you want to cut rather than bend the sheet, start by making a slightly deeper score than you would use for a bend and fold the sheet by more than 90° when bending. By working the sheet back and forth a few times, you will find that it is easy to make a clean break. I use a steel rule and a carpet knife (Stanley or similar) to score the sheet. **Photo 44** shows a sheet ready for bending into a complicated 3D structure for a home made enclosure. The sheet was marked with a permanent marker and then scored using the steel rule as a guide.

I bought a small sheet of 1mm aluminium from the friendly local sheet metal shop. This was large enough to make new front and rear panels for my enclosure and just enough spare to make a third panel in the likely event that I make a mess of the first attempt. The price was a remarkably cheap €2.

Layout

A large rectangular hole will be required for the 4x3 keypad. A rotary band switch will not be required because band switching is controlled by the PIC processor. I spent a few minutes drawing up a sketch of my proposed front panel layout, shown in **Figure 87**. There is some scope for making changes and plenty of room for adding a few more knobs or switches if this becomes necessary. The three rotary controls on the left are provisionally assigned to AF gain, manual IF gain and either RF drive or AF filter select. The 5 miniature switches on the right will be for SSB/CW, USB/LSB (this may end up as normal/reverse), AGC on/off, RF amplifier and attenuator. The area marked 'blank space' is reserved for any extra controls that may be added later.

Some front panels are designed to be aesthetically pleasing; others are only designed to be functional. Many of the larger items on the front panel will be held in place by mounting screws. If items like the LCD display or the keypad are screwed directly to the front panel, the screw heads will be visible on the front panel. **Photo 45** shows a typical example from the DDS signal generator project (March 2010). The LCD mounting screws are clearly visible. Using a mounting bezel for the LCD module and S-meter will

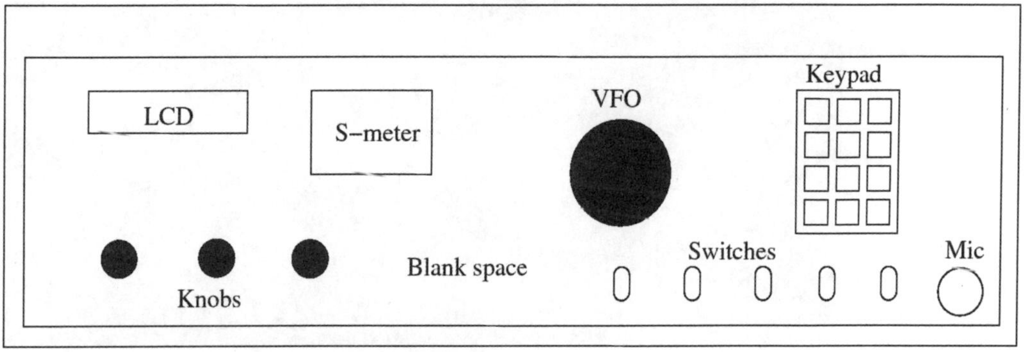

FIGURE 87: Draft layout of transceiver front panel.

PHOTO 45: Screw heads for things like displays are not always attractive.

result in a much tidier front panel. However, I have found that it is difficult or impossible to find suitable bezels for the most common types of LCD modules and keypads. One approach to this problem is to use a double front panel where the outer panel only has the minimum possible number of holes. Large items like the frequency display, S-meter, keypad and the optical encoder for the VFO are screwed to the inside pane only. If the mounting screws and holes are countersunk, it may be possible to sandwich them between the inner and outer panels so that screws and other mounting hardware are concealed.

Most of my projects are in enclosures with a simple single-layer front panel. I try to strike a reasonable balance between function and aesthetics by keeping most of the mounting screws hidden from view. Sometimes it will be necessary to have some mounting screws visible. I think it is acceptable to have the panel mounting screws visible at the corners of the panel. You can always use some nice new hex (Allen) or Torx head screws to mount the panel. If this is not acceptable, the screw heads can be concealed under a SNAP-CAP or a similar type of plastic cover.

Mounting Methods

I have tried many different mounting methods over the years. Many of these efforts have failed because none of the usual adhesives will

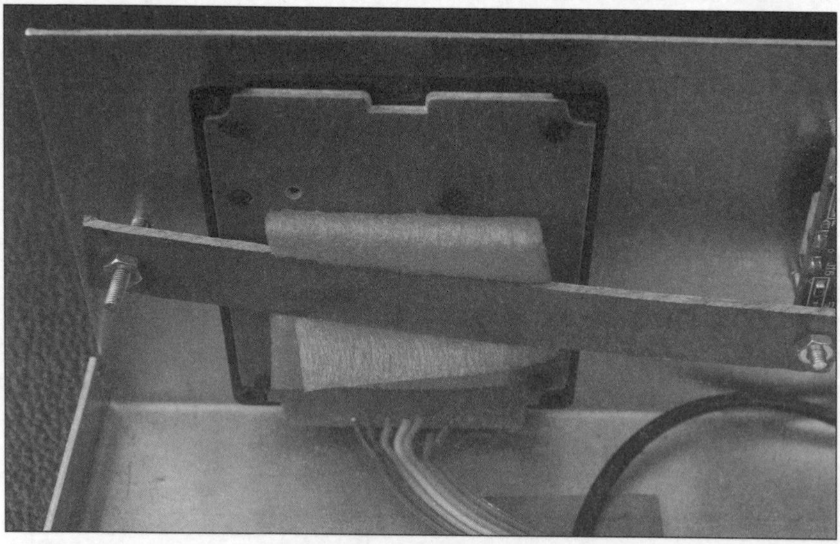

PHOTO 46: Using a strap and foam pad to secure a keypad means fewer screws.

bond well with aluminium. Super glue, epoxy resin, double sided tape, foam sticky pads and various types of tacky putty have all been tried with only limited success. For example, the moving-coil meter on the front of my directional wattmeter is held in place by a few blobs of Blu-Tack. This is great stuff for sticking my paddle key to the bench, but it doesn't work well with aluminium or some plastics. I have had to replace the Blu-Tack and remount the meter on several occasions.

One way to minimise the number of screws required for front panel mounting is to use a clamp to keep large items in place. **Photo 46** shows how I mounted the keypad in the DDS signal generator using a clamp made from a strip of PCB laminate. By using one of the screws for the LCD module as an anchor for one end of the clamp, it was possible to mount the keypad using just one extra screw instead of the four screws required by the normal mounting method. The foam pad protects the back of the keypad. This method also works well with panel mounting meters. Clamping an LCD display can cause problems. Even moderate clamp pressure may interfere with the operation of the display. Higher pressure might even crack the glass panel.

Cutting Holes

The three rectangular holes were cut out using a nibbling tool. It is quite easy to cut 1mm sheet. Slightly thicker sheet will make a more rigid front panel, but make sure you don't exceed the capacity of the nibbling tool. Mine is rated for 1.5mm. Mark out the rectangle using a permanent marker – a CD/DVD pen is ideal. Drill a large hole in the centre of the section to be removed. My nibbler will fit through an 8mm hole. **Photo 47** shows how the hole was cut for the frequency entry keypad. The nibbler only takes a few millimetres with each bite, so it will take some time make the hole. Use the nibbler from the back of the panel because the steel plate on the top of the nibbling tool will leave marks on the panel. Keeping damage to a minimum at this stage will reduce the amount of repair work that will be required at the priming and painting stage.

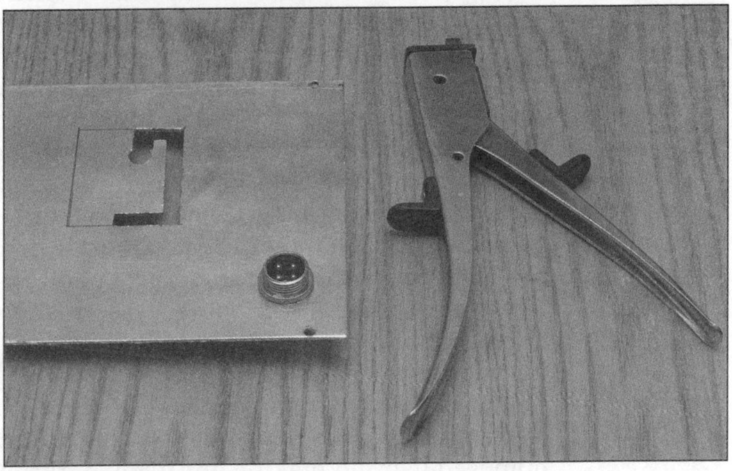

PHOTO 47: Starting with an 8mm hole, the sheet metal nibbler can cut nice straight edges for rectangular apertures.

After all of the front panel holes were cut or drilled, the rough edges of the rectangular holes were cleaned up using a flat file. Round holes can be

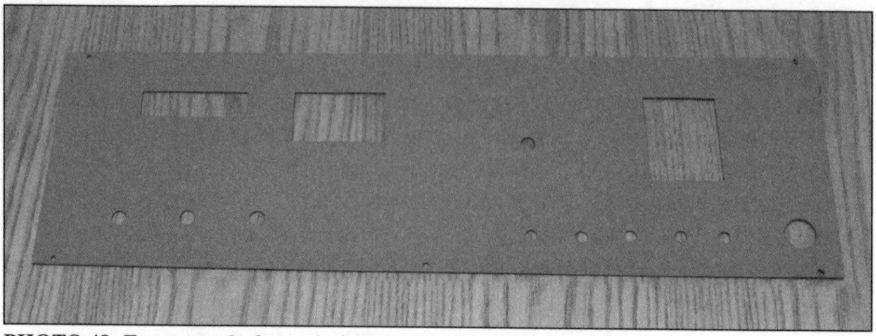

PHOTO 48: Front panel after priming.

cleaned up by using a larger drill bit as a hand tool. I used an 8mm drill to clean up the edges of the 6mm switch holes. DON'T use a power drill for this job – it is all too easy to go all the way through the panel. The panel was sanded using 150 grade sandpaper, As I didn't have any finer grade of wet-or-dry paper for the final sanding, I scrubbed the panel using coarse steel wool and finally with fine steel wool and warm soapy water. After rinsing with clean water, this left a spotlessly clean surface free from any marks or deep scratches. Make sure you don't touch the front panel before applying the primer. It is hard enough to get paint to stick to aluminium even without greasy fingerprints on it.

Painting

There are special etching primers for use on the aluminium panels used on caravans, aircraft etc. As I didn't have anything more suitable, I just used ordinary 'auto spray' grey primer from an aerosol can. I found that the best method is to apply several light coats with a couple of minutes between each coat. This seems to prevent any adverse reaction between the primer and the metal. The panel should be laid flat to avoid runs. While it would be best to prime and paint the panel in a warm, dust-free indoor environment, aerosol paints use a lot of thinner, which will leave your home smelling like a car paint shop for days. It is also very easy to get overspray on the surrounding area. I decided that it would be wiser to paint the panel in a cold outdoor shed, even if it would take much longer to dry and the panel gains a few spiders stuck in the paint.

Read the instructions on the can before using spray paint. Always spray in a well ventilated area. Wear a mask and avoid breathing the paint fumes. Don't spray near a heater or naked flame. The thinner and propellant used in aerosol paints is extremely flammable. The type of paint sold in car accessory shops and hardware stores is reasonably safe to use provided you follow the manufacturers instructions.

Photo 48 shows the panel with several coats of primer. Due to the fact that I used a nice new sheet of aluminium and I somehow managed to avoid making any deep marks or scratches on the front surface, the end result is a pristine panel that is ready for a very light rub of very fine (400-600 grade) wet sandpaper. Make sure you don't rub through to the metal or you will have to give the panel another few coats of primer before you can proceed to the painting stage. If there are any scratches or pits visible through the primer, they can be filled with a thin film of stopper putty before sanding. Allow plenty of time for the stopper to dry before final sanding.

Painting the panel should be relatively easy because you are now spraying on a nice clean primer base. Lay the panel flat and apply several light coats before applying a thicker final coat. Keep the spray can at a constant distance from the panel. Don't stop moving while the spray nozzle is pressed, or this will cause a localised build-up that may cause a run. Try to get perfectly even coverage by using slightly overlapping strokes of spray.

Labels

Once the paint is dry, you will need to find a way of applying labels to the controls. In the past, this was merely an aesthetic requirement. However, proper labelling of controls is now a requirement under the EI (Irish) licence conditions **[32]**. There are several ways of applying front panel labels. For non-artistic types like me, drawing a label freehand is out of the question. Labels can be painted on through a stencil, which will give more consistent results. Stick-on labels or transfers will often give better results, but the commonly available types have a solid colour background. This means your labels will have black text on a white background or vice-versa. A recent trip to the stationers and office supply shops revealed that the current favourite is gold letters on a solid background, yuck! Dymo tape or similar types of embossed tape can give good results. This method is subject to some of the same limitations as stick-on letters. However, embossed tape can give good results, especially when the panel colour is a perfect match to the tape colour. Black tape with white lettering works quite well on a black panel.

Some of the methods that we have previously used to make PCBs can also be applied to panel labelling. The toner transfer method can be used to apply black lettering directly to a light coloured panel (see Appendix 1). This method requires quite a lot of heat and pressure, so make sure your nice new paint job is well cured before you try it.

Screen printing is another way of making nice labels for your homebrew equipment. The process is very similar to the photographic method for producing PCBs. A very fine mesh screen is coated with a photo-sensitive emulsion. This is exposed to a UV light source through a positive or negative artwork, depending on the type of emulsion used. The screen is then developed (washed) and the paint or ink is applied to the panel through the screen using a wiper called a squeegee. I have used this method to apply etch resistant acrylic paint to PCB laminate.

I have found that dry rub-on transfers are still the best and most reliable method of applying panel labels. Letraset transfers are available directly from the manufacturer or from on-line sites like

PHOTO 49: Letraset-type letters can give good, durable results – this panel has been in use for nearly 20 years.

eBay. The letters are transferred from the transparent sheet using the wooden tool now supplied with each pack. **Photo 49** shows part of the front panel of my old transceiver. After almost 20 years of abuse, the lettering is still in very good condition. The lettering is covered with several coats of clear lacquer. Take great care when applying the first two coats of lacquer. The plastic letters are easily melted by the powerful solvents in the paint thinner. The first two coats should be a very light 'dust coat'. This will isolate the letters from the solvents in the final heavier coats.

Photo 50 shows the new enclosure ready for installing the modules we have built over the last year. I have screwed two sheets of PCB laminate to the bottom of the enclosure. This provides an excellent common ground plane for all the modules. Larger modules like the band pass filter unit will be mounted on this ground plane using M3 screws and nuts. Some of the smaller modules can be mounted vertically by soldering the PCB directly to the enclosure ground plane.

[32] http://tinyurl.com/bwykgxx

PHOTO 50: Enclosure ready to accept the transceiver modules. Note PCB laminate on the base.

14

Final boxing-up of the HF transceiver

THE MODULES built so far have now been installed in a 19in case. The (almost) finished project can be seen in **Photo 51**. I say 'almost' because a project like this is never really finished. There will surely be a few more tweaks to both the hardware and Pinguino firmware.

Hardware changes are relatively easy because of the modular nature of the project. Software changes are even easier because the PIC microcontroller can be reprogrammed via a USB cable. On-air testing in mid December 2011 has resulted in many favourable reports on the quality of the transmitted signal. The receiver seems to be at least as good as the receiver in my previous rig. It may even be better in some respects.

Figure 81 shows a block diagram of the transceiver. Some of the blocks were built specifically for this project, others were based on previous projects.

If you use the Class AB MOSFET PA (page 88 Figure 33), you will

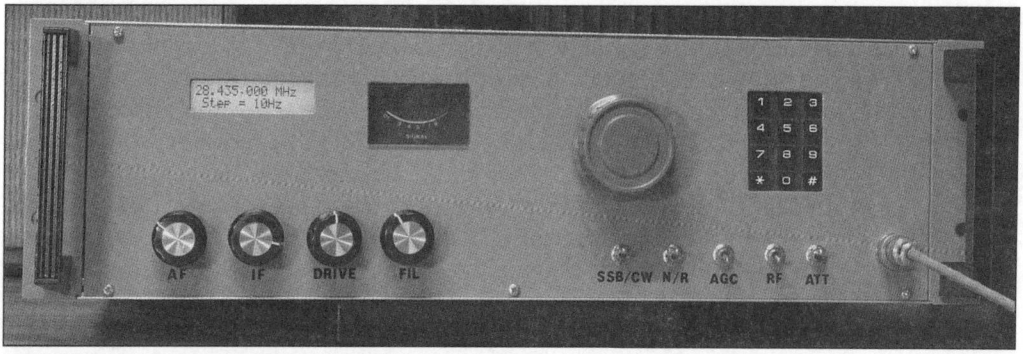

PHOTO 51: The (nearly) finished transceiver installed in the case .

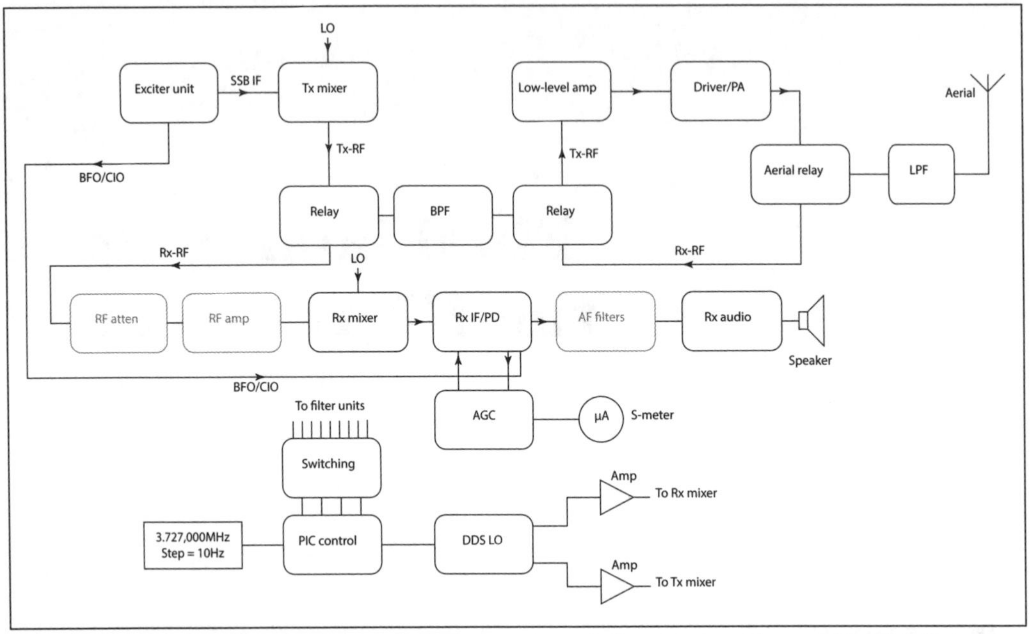

FIGURE 88: Block diagram of the transceiver. The grey boxes indicate projects that could be added later in the transceiver's life.

need the 1W bipolar amplifier in **Figure 82**.

This is the great advantage of modular construction: if you aren't happy with the performance of any individual stage, you can just rip it out and try something else.

As can be seen from **Figure 88,** the band pass filter (BPF) unit, aerial relay and low pass filter (LPF) unit are bidirectional and common to the transmit and receive signal paths. The transmitter circuit blocks are shown at the top of the diagram and are generally isolated from the receive circuit blocks except for the transmit SSB exciter unit, which also supplies the BFO/CIO signal to the receiver product detector. The BPF unit is switched between the receive and transmit sections by a pair of miniature SPDT relays (I used Maplin type N15AW). These are small, fast and quiet in operation, but not completely silent like the miniature gas-filled relays used in my old transceiver. **Figure 89** shows the configuration of the relays. I used a pair of 100nF capacitors to decouple the 12V DC supply to each of the relay coils. This is to prevent RF signals getting in or out of the BPF unit via stray capacitance between the relay coil and contacts.

The DDS local oscillator (Chapters 8-10) produces 300-400mV pk-pk into a 50Ω load. This must be increased to a suitable level and fed to the separate receive and transmit mixers. I used a two stage amplifier between the DDS and each mixer. Because the Rx/Tx LO signal is split between the first and second amplifier stages, three amplifiers are required. The configuration is: DDS LPF, first LO amplifier, passive resistive power divider and a pair of identical second LO amplifiers. All of the amplifiers are designed for 50Ω I/O

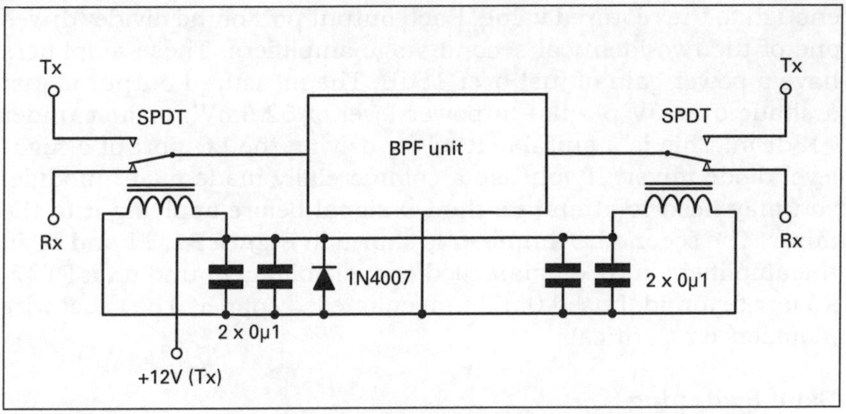

FIGURE 89: Two relays switch the BPF unit between the transmitter and receiver.

impedance. **Figure 90** shows the DDS LPF, first amplifier and the resistive 50Ω divider. The current configuration of the DDS LPF has a cutoff frequency of 41MHz. This allows band coverage from LF to 6m when a 10.7MHz IF is used. I started with a design for a 7th order 0.1dB Chebyshev LPF and then tweaked the component values to the nearest available standard values. The 256nH inductors are 7T of 1mm enamelled copper wire on a T50-6 toroid core. The 280nH inductor is also 7T on a T50-6, using closer turn spacing so that only 60% of the core is occupied. The first amplifier stage has about 12dB of gain. The power loss to each output port of the divider is 6dB, giving an overall gain of just 6dB. The correct resistor value for each of the three resistors in the divider is 50/3=16.6666Ω. A parallel pair of 33Ω resistors gives a value of 16.5Ω, which is close

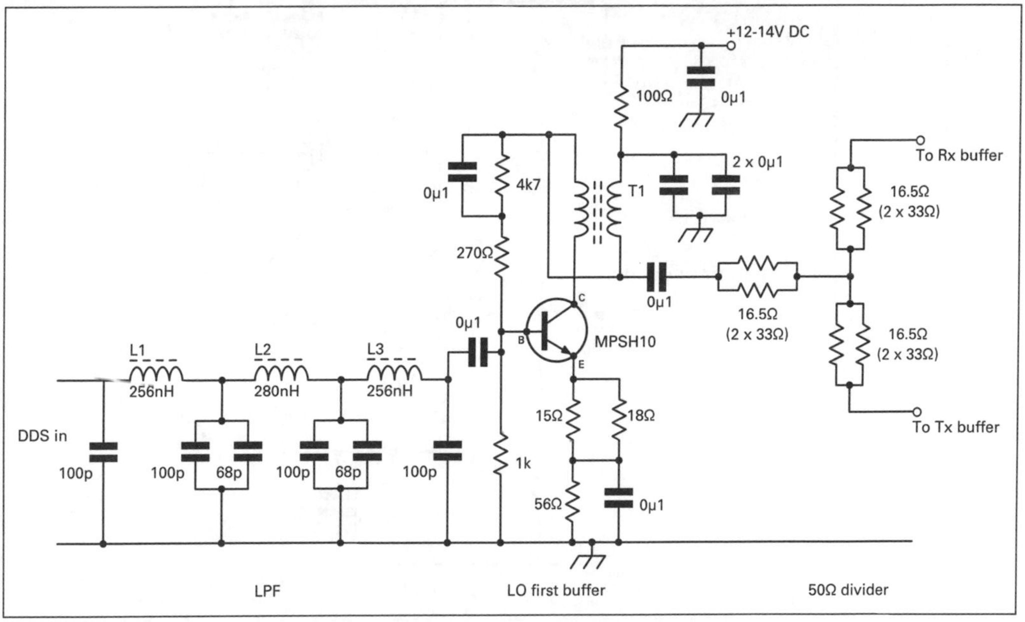

FIGURE 90: DDS LPF, first amplifier and the resistive 50Ω divider.

enough to the required value. Each output port of the divider drives one of the two identical second stage amplifiers. These amplifiers have a power gain of just over 15dB. The measured output is just a shade over 5V pk-pk – a power level of 62.5mW, or just under +18dBm. This is a suitable level for driving the LO port of a high-level diode mixer. If you use a commercially made mixer module, you may need to attenuate the LO signal before applying it to the mixer. The second LO amplifier is shown in **Figure 91**. T1 and T2 in the amplifiers is 6T of enamelled copper, bifilar wound on a FT37-43 ferrite toroid. I used 0.375mm enamelled copper. The exact wire diameter isn't critical.

Band Switching

A standard 4x3 keypad is used for direct frequency entry. To avoid the need for a separate band switch or band up/down buttons, I decided to use the keypad for band selection. Normally, the first digit of a new frequency entered would be 0, 1, 2 or 5, or possibly 7 if 4m is included. There should be no circumstances where the first digit would be 9, even if using all the amateur bands up to the microwave regions. So I have chosen the '9' key as the band select

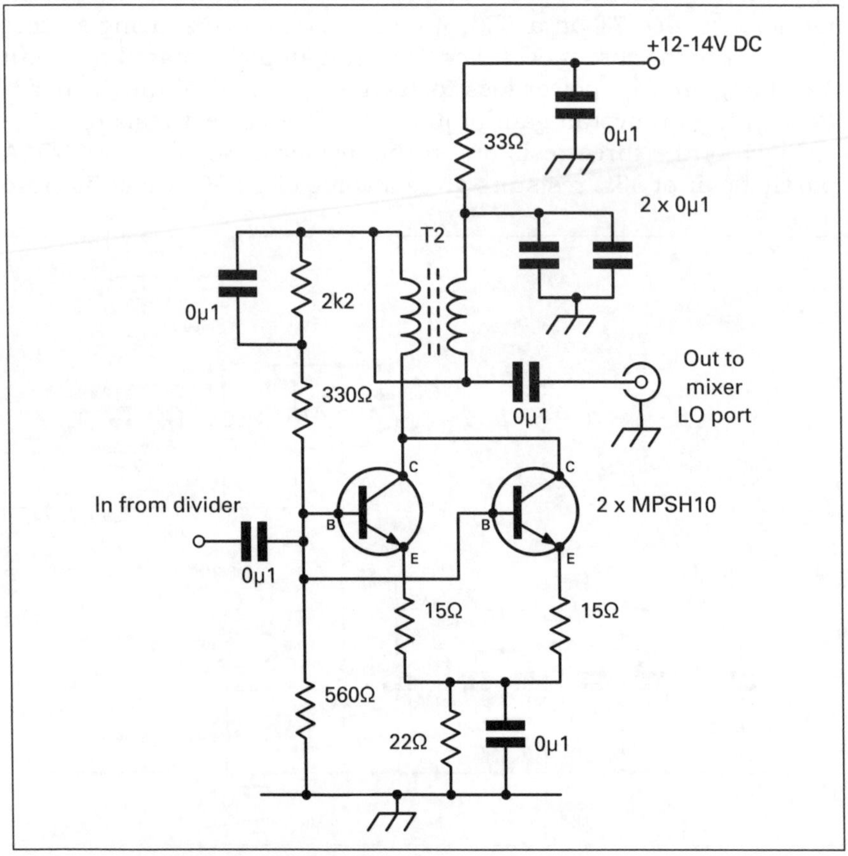

FIGURE 91: Second LO buffer amplifier.

button. I could also have used the '*' or '#' keys, but these have already been assigned for other purposes. This simple line of code is used to initiate a band change:

```
if(key_char=='9'){select_band();}
```

The select_band function presents a simple message that invites the user to select the required band.

```
clear();
display_string("Select Band:");
line2();
display_string("0=160m 9=6m");
```

These functions were written for our previous AD9851 DDS project. It is a very simple task to read the keypad using the get_key() function then make a decision based on the returned value. For example: it the '1' key is pressed:

```
case '1': { change_band(band_80);
dial=3727000; break; }
```

This sets the switches for the BPF and LPF units to the 80m band and assigns a new dial frequency of 3.727MHz.

The local oscillator frequency will be set to signal+IF for the lower frequency bands below 10.7MHz and signal–IF for the HF bands above 10.7MHz. To calculate the correct LO frequency, the control software checks to see if the dial frequency is greater than 10.7MHz:

```
if(dial>10700000) { offset=~OFFSET _N-1; }
```

A value of 10697000 has been assigned to the constant OFFSET_N. This is the normal (as opposed to reverse sideband) IF offset in Hz. The variable 'offset' is added to the dial frequency before it is loaded into the DDS. Note that addition is used regardless of whether we want to add or subtract the IF offset. If the dial frequency is greater than 10700000, the variable 'offset' is assigned a new value, which is the normal offset value negated (made negative) -10697000. Obviously adding a negative value to the dial frequency gives a negative offset to the LO frequency. One advantage of this scheme is that there is no need to change the USB/LSB carrier oscillator frequency when changing bands. If the carrier frequency is set for USB on the HF bands, the sideband inversion caused by using signal+IF LO injection on the LF bands means that an LSB signal will be generated. For this reason, the sideband selection switch is labelled normal/ reverse instead of USB/LSB. At this stage, the control system has no way of checking the normal/reverse switch position. This results in a 3kHz error in the displayed frequency when the sideband select switch is in the reverse position. As there are still several unassigned I/O pins on the Pinguino PIC board, this problem will be easy to correct. It will also be useful to sense the state of the transmit/receive switching to facilitate split frequency and cross-band operation. These features will be included in future firmware updates.

The current state of the firmware allows VFO tuning using a rotary encoder. VFO step size is selected using a push switch. The switch is part of the mechanical rotary encoder unit. This may be moved to a separate button if/when I replace the cheap mechanical encoder with a higher quality optical encoder. I could also move the step selection into a menu selected by the keypad. Currently available tuning steps are 1Hz, 10Hz, 100Hz and 1kHz. The VFO frequency can be directly entered on the keypad. It is possible to enter the entire frequency down to the last Hz, or you can take a shortcut by typing the first digits and then pressing the # key. Pressing 28# will load a frequency of 28.000,000MHz.

During a frequency entry sequence, pressing * will abort the sequence, leaving the VFO frequency unchanged. This is very useful when you accidentally press the wrong button.

I have built and installed filters for 160, 80, 40, 17, 10 and 6m. New filters can be added as time and the necessary components are available. Performance of the transceiver on the installed bands is very good. All spurious signals are generally better than 60-70dB below the wanted output. Spurious signals are undetectable on 160, 80 and 40m. This is probably because the LO, IF and IF image frequencies are all above the cutoff frequency of the 7th order LPF. The combination of this filter and the 4th order BPF gives huge attenuation of signals above the 10.7MHz IF.

The choice of 10.7MHz as the IF will probably cause some problems when I add the remaining HF bands. 2x10.7 = 21.4, which is likely to cause a spur at unacceptable levels on 15m. Balanced mixers give good suppression of the IF input and on odd harmonics, but very little suppression of 2f and other even harmonics. The 10.7MHz IF could also cause a spur on 30m, but at least this will be reduced to a large degree by the balanced mixer and by the BPF because of the frequency difference of 550kHz. The ideal solution would be to change the IF to a more suitable frequency. The other traditional choice of 9MHz could also lead to problems (2x9=18). I have a large collection of high quality 10MHz crystals and some cheap 9.216MHz crystals. Neither of these are an ideal choice. A couple of homebrew ladder filters at a frequencyaround 8.6MHz should give reasonable results. Options that come to mind are the standard crystal frequencies of 8.192MHz and 8.867MHz. This might be a project for the future...

Testing

The transceiver has been tested on all available bands. The very first CQ call on 17m brought an immediate response from VE3QT in Ottawa. Signals were 59 both ways with very favourable comments on the signal quality. A call on 10m resulted in several contacts with North American stations. The HF contacts were made without the benefit of a receive RF amplifier. Sensitivity is good enough to hear a substantial increase in receiver noise when switching from a 50Ω dummy load at room temperature to the aerial. I used a centre-fed doublet 40m long, 12m above ground for the tests. Conditions were very poor on 6m in midwinter. I had a chat with EI8GQ to confirm the rig was working on this band. Many thanks to Gary for helping with the tests. I have made many contacts on 160, 80 and 40m. Reports on the transmitted signal have been very good. The receiver seems very well behaved, with no sign of any problems, even at night on 40m.

You can see in Photo 51 that the front panel is fitted with a rotary control for transmit RF drive. This control isn't implemented yet. I'm not sure how I will control the transmitter output without introducing some distortion. A simple switchable attenuator is probably the best option. Attenuation of 0, 3 and 10dB would give 10W, 5W, and 1W from the Class A PA. Variable-gain amplifiers or attenuators would give more flexible control of output level, but they would also introduce the risk of increased distortion.

15

A 200W MOSFET PA for 160-6m

I DECIDED TO use a pair of 50V MOSFETs. There are several standard devices to choose from. My previous MOSFET amplifiers were based on the ST SD2931-10, a 50V, 150W, 175MHz MOSFET. The MRF150, MRF151, VRF151 and Philips BLF177 are very similar devices with almost identical specifications. The new amplifier uses a push-pull pair of MRF151s. Any of the other 150W 50V devices should give similar performance. As with the previous MOSFET amplifiers, I will take a conservative approach to the design. The power supply voltage will be reduced to 40-42VDC and power output will be limited to 100W per device (200W for the pair).

Figure 92 shows the schematic of the MOSFET amplifier. This is

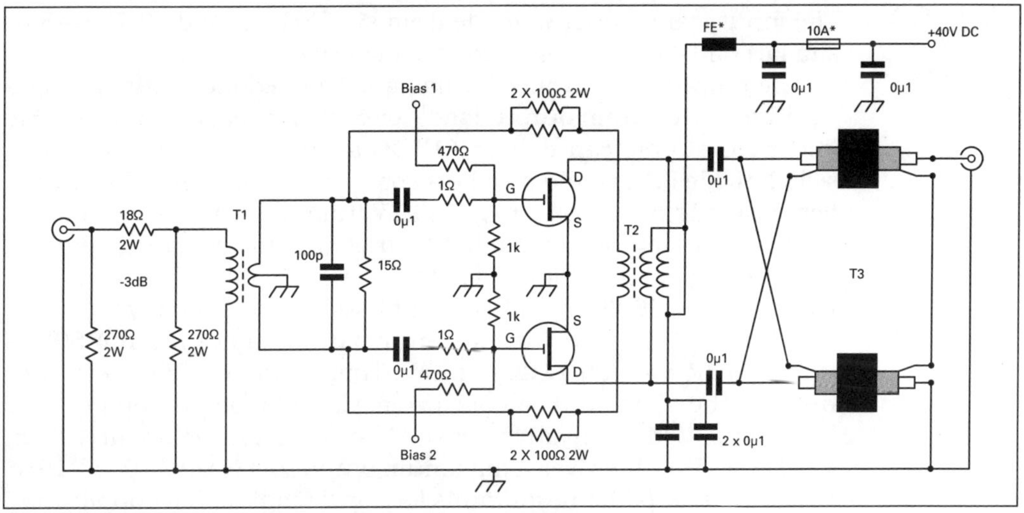

FIGURE 92: Circuit diagram of the 200W MOSFET amplifier.

the classic balanced pushpull configuration. The input transformer is a conventional broadband transformer with a three turn primary and a single turn secondary made from brass or copper tubing. The secondary is centre-tapped and the centre terminal is grounded to provide balanced push-pull drive to the FET gate terminals. The 3:1 turns ratio gives a 9:1 impedance ratio. The output transformer is a transmission line type, which is known to work well over a very wide bandwidth. This is a scaled-up version of the output transformer used in both of the PA designs for the HF rig. Remember that the MOSFET version worked well from LF up to 100MHz.

The resistive part of the input impedance of these FETs is around 1-3Ω at HF and VHF. At LF the input resistance is higher and the gate capacitance appears as a high value of series capacitive reactance. A combination of resistive input loading and negative feedback (NFB) will tend to reduce the input impedance at LF so that the input impedance is consistently low across a wide bandwidth. The reduced gain associated with resistive loading and NFB can be regarded as an advantage rather than a problem because RF MOSFETS usually have gain well in excess of 20dB. This is excessive for most RF power amplifier applications. Reducing gain to 13- 16dB will usually result in greater stability, better linearity and wider bandwidth.

The input resistance of the push-pull pair is around 2x2Ω =4Ω, or perhaps a bit higher on the lower bands. Each FET gate has a 1Ω resistor connected in series, giving a total gate-to-gate resistance of around 6Ω. The 9:1 transformer should therefore give a reasonably good match to 50Ω. The transformer was tested using a return-loss bridge and a 5.45Ω resistor as the secondary load. The resistor was made from a parallel combination of 10Ω and 12Ω surface mount chip resistors. Best bandwidth was achieved with a frequency compensation capacitor of more than 200pF across the 1 turn secondary. When the transformer was installed in the amplifier, the optimum value was found experimentally to be around 100pF. The input transformer is made from six ferrite toroids (FT37-43 or similar) on brass tubing or coax cable braid.

The output transformer T3 has a 4:1 impedance ratio, giving a 12.5Ω drain-to-drain output load – or 6.25Ω for each transistor. My linear DC supply can deliver 41VDC at very high current. A 42V switch mode PSU would also be a good power source for this amplifier. It should be possible to get 300W from this amplifier if you use a 48V or 50V supply. I wouldn't be prepared to accept the reduced safety margins. I would prefer to have a reliable 200W amplifier rather than a 300W amplifier that blows up occasionally.

With a 41V supply, the maximum saturated output into 50Ω is $(Vdd*4)^2/2R = (41x4)^2/100 = 269W$. This is not possible in practice because the maximum voltage swing will not reach 0V on negative peaks or 2*Vdd on positive peaks. Assuming Vsat is about 10% of VDD (or 4-5V), this gives a maximum output of $(37\times4)^2\div100 = 219W$. Driving the amplifier beyond this level will result in flat-topping and severe distortion. If you really need more than 200W, you can always

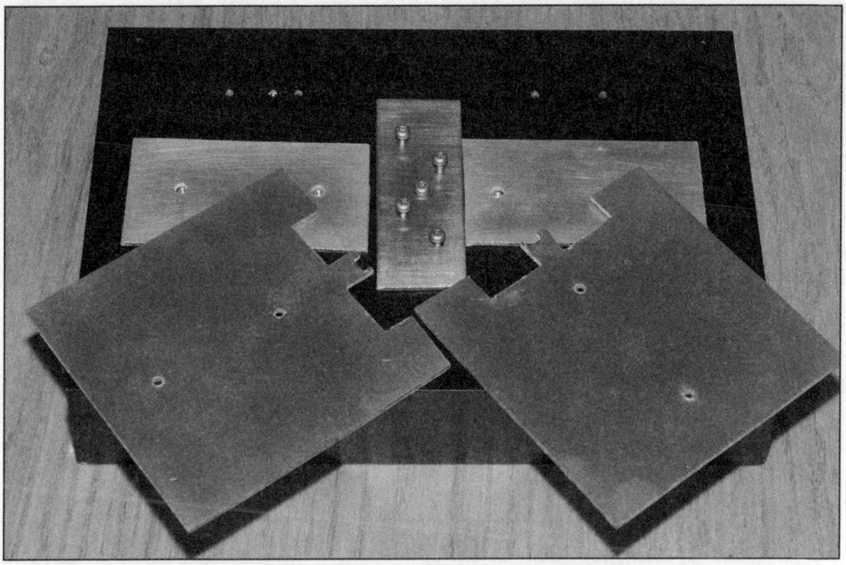

PHOTO 52: The main hardware elements of the 200W amplifier prior to assembly..

build two or more of these amplifiers and some power combiners.

Construction

The amplifier was built dead-bug style on PCB laminate. The transistors are mounted on a copper heat spreader that is screwed to a large heatsink. **Photo 52** shows the main hardware before the amplifier was assembled. The heat spreader is 2.5mm thick copper. This is thinner than recommended, but it was all that was available at the time. 4-6mm would be better. The copper was lapped flat on both sides using a sheet of wet or dry sandpaper and a thick sheet of plate glass. I used a thin film of ordinary white heatsink compound between the spreader and the heatsink. An even thinner film of high performance Arctic Silver compount was used between the FETs and the spreader. The output transformer is two lengths of Maplin XR16 screened audio cable. Each winding is four turns on a Maplin N90AB ferrite core. T2 is six turns made from three twisted lengths of 1mm enamelled copper (trifilar wound) of a Maplin N88AB ferrite

PHOTO 53: General view of the 200W amplifier.

PHOTO 54: The output end of the amplifier.

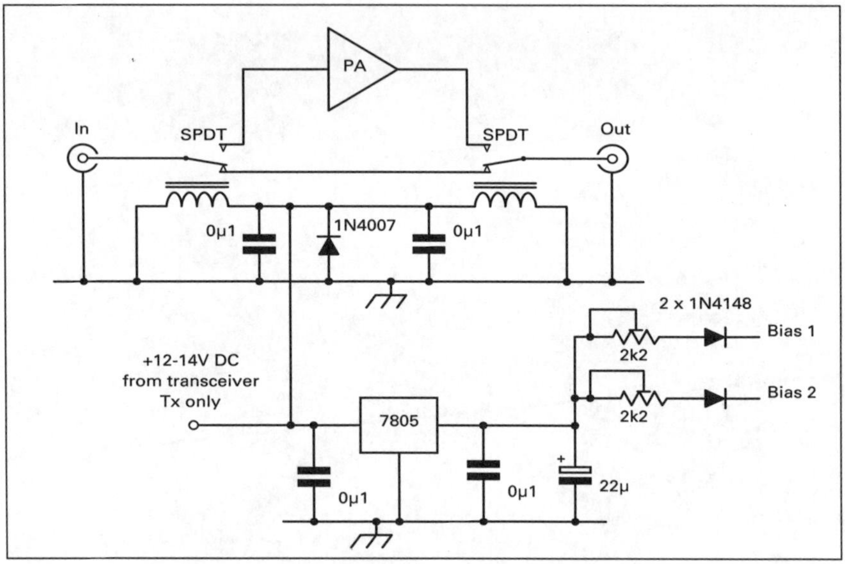

FIGURE 93: Bias and switching arrangements for the amplifier.

core. **Photo 53** and **Photo 54** show views of the finished amplifier.

Initial testing of the amplifier showed a power gain of 20dB, or 200W out for 2W in. NFB was increased slightly. A 15Ω resistor was placed across the secondary of T1. This improved input return loss (lower SWR), particularly on the lower bands. Finally, a 3dB attenuator was added to the input circuit. It now takes 5W of drive for 200W out. Low level gain measurements show a flat response from 1.7MHz to 60MHz. High level tests show that there is a slight roll-off at 6m, where output is 150W out for 5W in. LF to 30MHz gain is perfectly flat. Second harmonic output is well suppressed at -30 to -40dBc. Third harmonic output is not so well suppressed, at only -15 to -20dBc. A 7th order LPF was used for all on-air testing. Transmit quality is very good with no sign of adjacent channel splatter. **Figure 93** shows the bias and Rx/Tx switching.

16

A 400W HF PA

THE TYPE OF output transformer used in most broadband RF power amplifiers has a very small number of turns on each winding. In many cases, the primary winding (FET drain or transistor collector) consists of just a single turn made from metal tubing or coax braid. The primary is often centre tapped for push-pull amplifiers. The transformer impedance ratio is the square of the turns ratio. A 1 turn primary and 2 turn secondary gives an impedance ratio of 1:4, 1 and 3 turns give 1:9 and so on. Because of the relatively low supply voltage of 37V, I chose to use a 1 and 4 turn transformer in the 400W PA, **Photo 55**. Allowing for a FET saturation voltage (V_{sat}) of about 4-5V, this gives a maximum output voltage of (32 x 2 x 4) = 256V. This indicates a maximum output of 655W and good linearity at the specified output of 400W. The drainto- drain load impedance is 50/16=3.12Ω, which is quite a bit lower than recommended. Increasing the supply to 42V calls for

PHOTO 55: General view of the reworked 400W amplifier.

PHOTO 56: Output transformer components.
Top to bottom: two Maplin N90AB toroids, end plates made from made from single sided PCB laminate and Westflex 103 outer conductors.

a lower transformer impedance ratio. As we can only use increments of one full turn, the next available option is to use three turns on the secondary. This gives an impedance ratio of 1:9 and a drain-to-drain load of 5.555Ω. The SD2931-10 specifies an optimum RL of 6.8Ω for 30MHz at 50V, so 5.555Ω should be ideal for a 42V supply. Maximum output is $(V_{cc} - V_{sat})^2 / 2R$ = 123W per transistor (or 492W total for the push-pull arrangement of four transistors).

Testing the small 16:1 input transformer using a signal generator, return loss bridge and a 3.333Ω load made from three paralleled 10Ω 1% tolerance resistors showed that the transformer is well behaved from LF to above 30MHz. There was no need for any frequency compensating capacitors. The high input capacitance of the FETs resulted in a slightly overcompensated response, with a small peak just above 30MHz. I decided to turn my attention to the output circuit.

As the original output transformer is seam soldered along the entire length of both of the PCB end plates, it would be difficult to remove it from the amplifier. I decided it would probably be easier to make a new transformer for the RF bandwidth tests. The transformer is a 'conventional' type, with a single turn centre tapped primary made from Westflex 103 coax braid and end plates made from single sided PCB laminate. The transformer core is a pair of large ferrite toroids (Maplin N90AB). The transformer components are shown in **Photo 56**. For the first test, the transformer was terminated by a 3.333Ω resistor (as described previously). The four turn winding was connected to a 50Ω return loss bridge (RLB). The generator and detector ports of the bridge were connected to the tracking generator and RF inputs of my spectrum analyser. Return loss (RL) was very

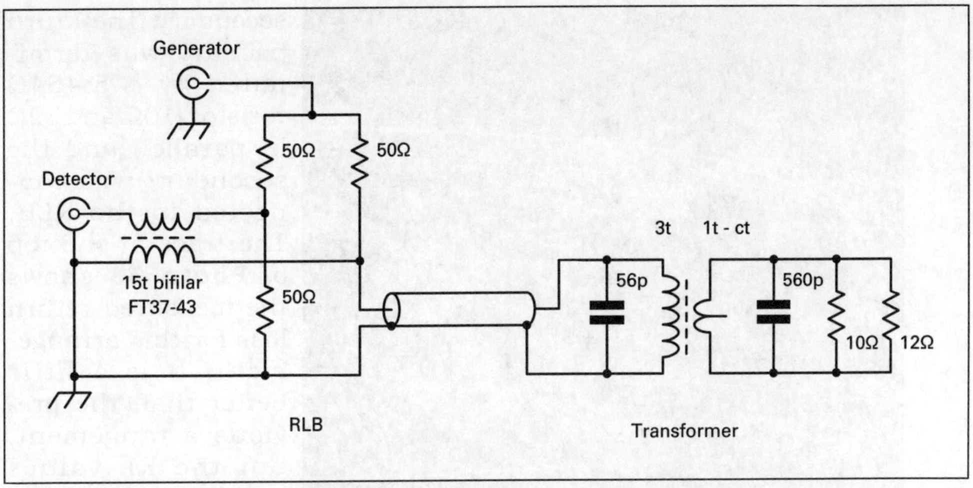

FIGURE 94: Circuit diagram of the return loss bridge and transformer under test.

high at the LF end of the spectrum. This indicates a good match and a very low SWR. At 14MHz, RL was around 10dB (SWR 2:1). At 30MHz, return loss was only about 5dB (SWR 3.6:1). This probably accounts for the poor performance on 10m. **Figure 94** shows my RLB and transformer test rig. The RLB is made from three 50Ω resistors and a simple balun. I used parallel pairs of 100Ω resistors for each arm of the bridge and a FT37-43 toroid with 15T of bifilar wound enamelled copper as the balun. The transformer test rig is shown in **Photo 57**.

For the second test, I removed one turn from the transformer

PHOTO 57: Transformer test rig of Figure 93. Transformer on left, return loss bridge on right.

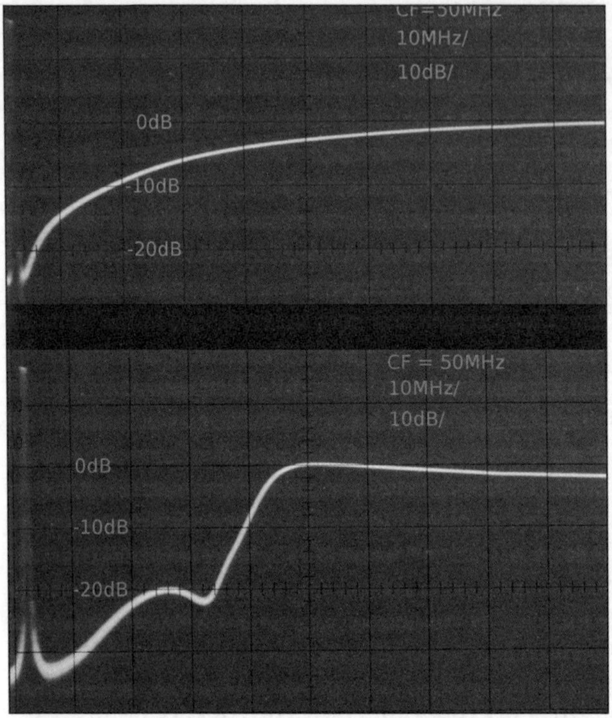

PHOTO 58: Return loss measurements, 0-100MHz. Top: 1T primary terminated in 5.454Ω, 3T secondary to RLB; performance poor. Bottom: after adding compensating capacitors, the improvement is quite dramatic between 7 and 30MHz.

secondary. The 1 turn primary was terminated by a 5.454Ω resistor (10Ω and 12Ω in parallel) and the secondary was connected to the RLB. The trace at the top of **Photo 58** shows the measured return loss for this arrangement. It is a little better than the previous arrangement, but the RL values around 30MHz are still only around 6dB (SWR 3:1). The mismatch is mostly due to leakage inductance in the transformer windings. Measuring the inductance of the three turn secondary winding on my LCR meter shows an inductance of just over 40μH. This is more than sufficient to cover the 160m band. Inductive reactance at 1.8MHz is $2 \times \pi \times 1.8 \times 40 = 452\Omega$. This satisfies the requirement that the winding reactance is several times greater than the load impedance of 50Ω.

I soldered a small strip of copper foil (stripped from double screened coax) across the open end of the single turn primary. This makes a zero ohm termination (a perfect short). In a perfect transformer, the secondary resistance and reactance would also be zero. In a real world transformer, there are always some losses in the windings and in the core. Due to imperfect coupling between primary and secondary, there will always be some leakage inductance. The secondary inductance with a shorted primary was measured at around 300nH. Several measurements were made at several test frequencies. The LCR meter shows 340nH, further tests at 30-40MHz using a signal generator, RLB, a 50Ω load and a handful of 1% tolerance capacitors showed an inductance of 290nH. Measuring primary inductance with a shorted secondary is difficult because of the small inductance values involved. The primary leakage inductance is assumed to be 300nH / 9 = 33nH. A QUCS **[33]** computer simulation of the transformer using these values of leakage inductance shows even worse performance than the real transformer in the test rig.

300nH is a quite insignificant 3.39Ω at 1.8MHz and a very substantial 56.5Ω at 30MHz. Adding similar amounts of capacitive reactance in series with the transformer windings results in a perfect match at 30MHz. However, this approach won't work for a broadband transformer. The solution is to add capacitance in parallel with the transformer windings. I did attempt to calculate series-to-parallel equivalent values for the shunt capacitors, but I found that although the calculated values gave good results in the QUCS model they were not so good in the real world transformer. I eventually decided to rely on empirical methods by gradually increasing the value of shunt capacitance until the best RL plot was observed. The primary and secondary shunt capacitor values were kept at a ratio of approximately 9:1. Best results were obtained with 500-600pF across the primary and 56pF across the three turn secondary. As it happens, these are typical values found in commercial 1.8-30MHz broadband amplifiers. The lower plot in Photo 58 shows the measured RL from LF to 100MHz. The improvement is quite dramatic at frequencies between 7 and 30MHz. RL is generally better than 30dB (SWR 1.07:1) from 1-7MHz and 20dB (SWR 1.2:1) up to 34MHz.

The output transformer of the 400W amplifier was modified by reducing the secondary winding to 3 turns, placing a 56pF 2kV ceramic across the secondary winding and a pair of capacitors across the single turn primary. I used 500V mica, 390pF in parallel with 100pF. To improve the source-ground connection of the centre pair of FETs, I added an extra screw through the PCB ground plane, copper heat spreader and heatsink. The self-tapping screw and the extra capacitors are circled in **Photo 59.** I also replaced the 2.2k bias adjustment pots with 1k multi-turn types. This makes bias adjustment much easier. The new bias adjustment circuits

PHOTO 59: Modified output transformer area. The extra capacitors and screw are circled.

PHOTO 60: The new bias adjustment pots.

can be seen in **Photo 60** and the amplifier schematic can be seen in **Figure 95**. To keep output power within safe limits, I have added a 4dB attenuator to the amplifier input. Without the attenuator, the modified amplifier produces over 400W on 10m and over 500W on the lower bands. With the attenuator installed and 10W of drive from the new transceiver, it gives 300W on 10m and 400W on the lower bands. This arrangement reduces the possibility of accidental overdriving. It also improves input return loss by 8dB. The attenuator resistors are 2W metal film types. For low duty modes like SSB and CW, the resistors run cool. If you run 400W on high duty cycle modes, you will need 3W resistors and a big fan to keep the heatsink cool.

[33] http://qucs.sourceforge.net

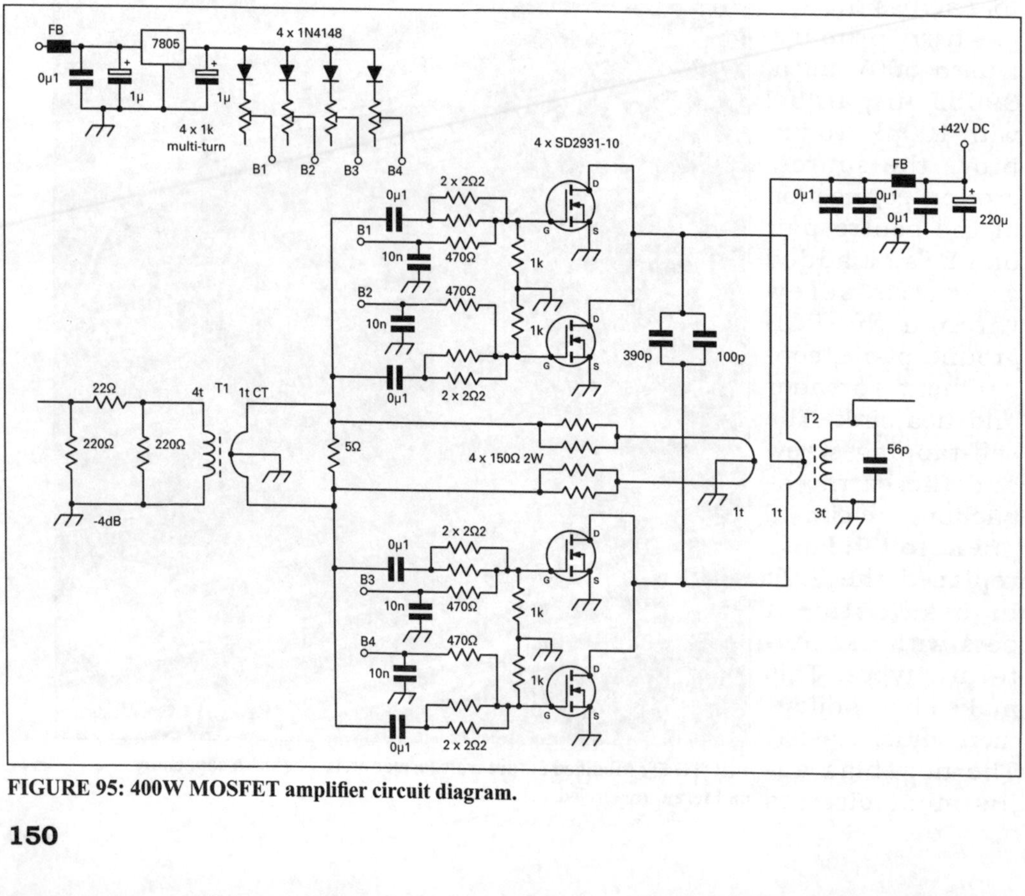

FIGURE 95: 400W MOSFET amplifier circuit diagram.

Appendix 1
Construction Methods

INTRODUCTION: I am a big fan of dead-bug construction and other similar point-to-point wiring methods. There are quite a few variations on this method. Some constructors use a hacksaw blade to cut the copper layer of PCB into a grid of isolated square or rectangular pads. The component leads are soldered to the pads and short lengths of wire are used to make connections between pads. **Photo 1.1** shows a simple circuit that was built using this method.

This example uses single-sided PCB. The components can be mounted on the same side of the board as the copper pads. If you drill holes for the component leads, the components can also be mounted on the opposite side of the board. If double-sided PCB is used, one side can be used as a continuous ground plane.

Another style of construction uses copper pads that are cut or punched from copper-clad PCB and glued to a copper ground plane. This method is known as 'Manhattan Construction'. Excellent Manhattan Construction tutorials can be found on K7QO's website **[2]**.

The methods described can produce attractive and functional circuits. I have found that the extra time and effort to cut out grids and pads can rarely be justified. Simple dead-bug construction is easier, quicker and just as reliable. To enhance the mechanical stability of a dead-bug circuit, some of the component leads can be soldered to PTFE or ceramic stand-off terminals.

Photo 1.2 shows a VHF balanced mixer that uses press-in PTFE stand-off terminals. Good old fashioned tag-strip terminals can also be used for this purpose. Very high value resistors can also be used as stand-off insulators. A 10MΩ resistor will

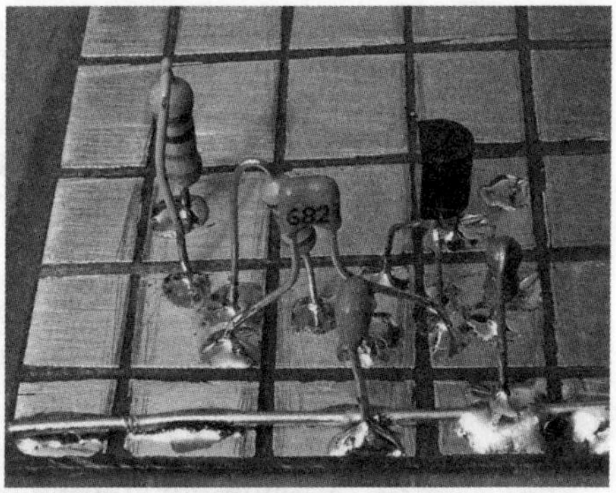

PHOTO 1.1: example of grid-based construction.

PHOTO 1.2: VHF balanced mixer that uses press-in PTFE stand-off terminals.

usually have little or no effect on the operation of your circuit. One end of the resistor is soldered to the copper ground plane.

PHOTO 1.3: A 1296MHz band-pass-filter made from cut and glued line.

UHF and microwave circuitry often uses transmission lines as the reactive elements of filters and matching networks. The transmission line sections can be etched on double-sided PCB or cut from single sided PCB and glued to a copper ground plane. Air-dielectric lines can be cut from sheet metal and mounted a few mm above the ground plane. The simplest and most common arrangement is called 'microstripline'. This type of circuitry can easily be integrated with the more conventional dead-bug techniques we are already familiar with. 50Ω transmission lines are easily made from standard fibreglass PCB. Assuming that the board is 1.6mm thick and has a relative permittivity of 5.0, a 50Ω line will be 2.77mm wide **[3]**. This line can be etched on 1.6mm G10 or similar double-sided PCB or cut from single-sided PCB and glued to a copper ground plane. The cut-and-glue approach is more forgiving when you make a mistake. **Photo 1.3** shows a 1296MHz (23cm) band-pass-filter made from cut and glued line. Each of the three lines is shorter than a quarter wavelength and is tuned to resonance by a small trimmer capacitor.

Making Your Own PCBs

The printed circuit board (PCB), also known as printed wiring board (PWB), is quite literally the backbone of modern electronics. The vast majority of electronic circuits are built on printed circuit boards. The most common type of PCB has one or more layers of copper foil bonded to an insulating substrate. The substrate is usually made from a composite material like fibreglass/epoxy or SRBP (synthetic resin bonded paper). The copper tracks and pads are formed by chemical etching of the copper foil.

Single sided and double sided copper clad board is readily available from most component suppliers. Fibreglass board is a bit more expensive than SRBP, but it has a few advantages.

Fibreglass board is much more durable than SRBP. It also has more predictable electrical properties, is more flame retardant and is easier to cut with a shear or a saw. SRBP board is less expensive than fibreglass, and it is also much easier to drill. SRBP can be difficult to cut because it cracks very easily.

Most PCBs are made by drawing the track layout on copper clad board using an etch resistant coating and immersing the board in an etchant solution. Bare copper is removed by the etchant, leaving only the copper protected by the etch resist. After etching, the board is rinsed with clean water, the etch resist is removed and the board is thoroughly cleaned using a mild abrasive. Finally, holes are drilled for the component leads. The components can be mounted on the opposite side of the board to the copper tracks or surface-mounted on the same side as the copper tracks.

SAFETY. Please take great care when working with fibreglass PCB. Cutting fibreglass with a hacksaw produces quite a lot of fine dust. At the very least, you should wear a dust mask and work in a well ventilated area. Wear protective goggles when you are drilling. Don't leave PCB etchant or developer trays within reach of children or animals. Don't rub or splash glass fibre or chemicals into your eyes. Wear rubber gloves when working with PCB chemicals.

ARTWORK. There are many different ways of printing the track layout onto the copper PCB. The most obvious way is to draw the tracks by hand with an etch-resist pen. Special pens are available for this purpose. I have found that some brands of laundry marker make excellent etch-resist pens. Staedtler fine laundry markers give particularly good results. Tesco own-brand CD/DVD pens also make good etch-resist pens but I haven't had much success with ordinary waterproof/permanent felt-tip pens.

I etched a small scrap of PCB to test the effectiveness of a Staedtler fine laundry marker, a Dalo etch-resist pen and a Tesco CD/DVD pen. **Photo 1.4** shows the etched board. All three pens gave acceptable results. A close inspection of the board with a magnifying glass shows that the laundry marker wins by a small margin. It has slightly better etch resistance than the Tesco CD pen and can draw thinner lines than the Dalo. The Tesco pens are good value at four for less than a pound.

A scrap of Veroboard can be used as a template for drilling holes in hand-drawn PCBs. I have found that the best approach is to drill the holes for IC pads first. The holes can then be used as a guide for drawing the IC pads and tracks.

CAD/CAM. Back in the bad old days, PCB artwork was laid out by hand using rub-on tape and pads. These days, most PCB layouts are created on computers running CAD (computer aided design) software. PCB design software ranges from simple PCB CAD programs to complete EDA (electronic design automation) packages. If you don't need advanced features like schematic capture, circuit simulation and auto-routing, you could use a standard drawing program to create PCB artwork. I use a free and open-source PCB CAD program with the rather unimaginative name of *PCB* **[4]**. *PCB* runs on Linux and other Unix-based systems. There are several PCB CAD programs available for Windows and MacOS, some of which are available as freeware. Simple PCB work is not very demanding in terms of memory and CPU cycles, so a fairly humble PC is adequate for this task.

PHOTO 1.4: A comparison of the effectiveness of three pens for use as an etch resist marker
*** Staedtler fine laundry marker**
*** Dalo etch-resist pen**
*** Tesco CD/DVD pen**

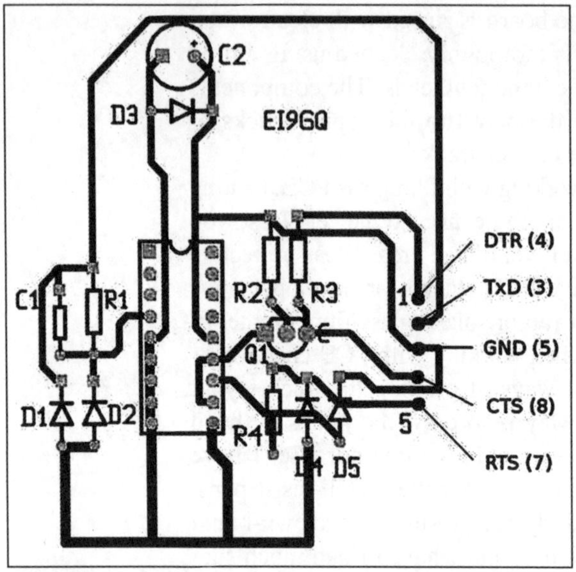

FIGURE 1.1: A typical layout for a simple single-sided PCB.

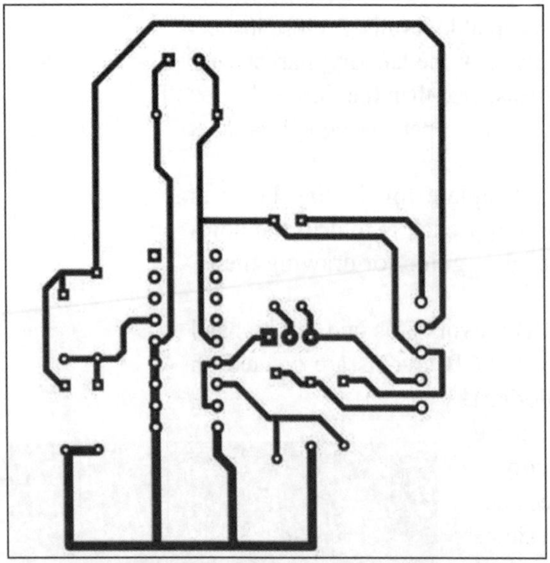

FIGURE 1.2: CAD-produced artwork used to create the PCB. Note that this is not to scale. Correctly scaled artwork can be downloaded from: http://homepage.eircom.net/~ei9gq/art.html

Figure 1.1 shows a typical layout for a simple single-sided PCB. The different colours show the various layers of the PCB. Even this simple example has a silk screen layer for printing the component outlines and names on the component side of the board, a solder resist layer that we won't be using and the copper track layer at the bottom of the PCB. Before the advent of surface-mount technology, single-sided PCBs usually had the components mounted on the top of the board and the copper tracks on the opposite side. Life was simple, the top was at the top, the bottom was at the bottom and everybody knew which was which. Boards using surface-mount technology (SMT) have the components mounted on the same side of the board as the copper tracks. Many PCBs have at least two copper layers.

Sometimes surface-mounted components will be mounted on both sides of the PCB or a mix of SMT and leaded components will be used. Complex boards are often made from a multilayer sandwich of copper and fibreglass. Such boards can have a ground plane layer, one or more power layers and several signal layers. Add a few more layers for solder mask and silk screen printing on both sides of the board and you will understand why some PCB CAD programs have such a steep learning curve.

Commercially made PCBs use plated through holes (PTH) to make connections between layers. Such boards would be very difficult to make using kitchen table technology. Most homebrew PCBs have a maximum of two copper layers. Home made double-sided boards use the component leads or short lengths of wire to make connections between the top and bottom copper layers.

Most PCB CAD software displays a top view of the PCB on the computer screen. This can be quite confusing when you are designing a conventional single-sided board. You are effectively drawing the bottom track layout while you are looking at the other side of the board. To avoid confusion, you can mark pin 1 of each IC or use IC pads which clearly identify pin 1. The standard DIL IC pads supplied with *PCB* (the software package) have a square pad for pin 1 and round pads for all other pins. This helps to avoid confusion. It is also a good idea to put some text on the board. Seen from above, text on the bottom copper layer will be a mirror image of the text that will be on the finished PCB.

PRINTING/PLOTTING. When the artwork is finished, we must find a way of getting the tracks and pads from the computer screen onto the copper clad PCB. There are several different ways of printing the etchresist pattern onto the board. Commercial PCB houses usually use a photographic process. The track pattern can be printed directly onto a transparent film using a plotter. Alternatively, the pattern can be printed on paper using a conventional plotter or printer. The pattern is then photographed with a special camera. The resulting photograph is used as a master artwork for exposing PCBs to ultra-violet (UV) light. The artwork is placed between special pre-coated photosensitive PCB and a UV light source. The photo-resist is softened by the UV light. When the board is developed, the exposed photo-resist will be washed away. Only the areas of photoresist that were shielded from the UV light by the opaque artwork will remain on the board.

Another method of producing PCBs uses silk-screen printing to print etch-resistant ink directly onto the copper board. The process of making the screen is quite similar to the photographic methods mentioned above. A fine mesh screen is coated with a UV sensitive emulsion. The screen is placed in close contact with the master artwork and exposed to UV light. The screen is developed by washing away the emulsion. The finished screen is used as a mask for spreading waterproof ink directly on the PCB surface. A rubber spreader called a squeegee is used to spread the ink. A single screen can be used to produce many PCBs in one session.

One of the most popular methods of homebrew PCB production uses toner from a laser printer or photocopier as the etch resist coating. The track pattern is printed onto a special transfer film using a standard laser printer or photocopier. The toner is then ironed onto the copper surface of the PCB using an ordinary clothes iron. Most laser printer toners are very waterproof and make excellent etch-resist. I have also had some success with plotting waterproof ink directly onto the PCB. I modified a Roland flat-bed plotter by removing the rubber pad from the plotter bed. As the rubber mat was approximately the same thickness as 1.6mm PCB, this left room to place the PCB directly on the plotter bed. I made quite a few boards using this method, some very successful, others not so good.

We will start by describing the process for making a PCB by the toner transfer method. The bottom copper layer from Figure 1.1 was used as the artwork. Remember that you will probably need to print a mirror image of your PCB layout on the transfer film. The process of ironing this image onto the board will mirror the image again and restore the original layout. Most CAD and graphics programs will have a mirror or flip function. Do a test print on plain paper before you commit yourself to using a sheet of expensive transfer paper. Any errors that you fail to spot at this stage could easily end up on the completed PCB! I know because I have made that mistake. Unfortunately, special PCB transfer films are far too expensive for the EI9GQ homebrew budget. A cheaper alternative will have to be found.

DISCLAIMER. Putting anything other than the manufacturer's recommended paper or film through your laser printer or photocopier could damage it. The printer/copier fuser unit is a high temperature roller used to melt the toner into the surface of photocopy paper. Some types of plastic OHP film or plastic coated paper could make a horrible mess of it. If in doubt, you can check the heat resistance of your paper by running a hot clothes iron over it. If you have any reservations about putting inkjet paper through your laser printer, then you should buy the 'proper' transfer film instead. I have had good results with TEC200 film. The most popular type of film is Press-N-Peel Blue which is available from Maplin and many other suppliers. See part 3 of G3XJP's excellent DSP transceiver project **[5]** for a good example of what can be achieved using toner transfer film. I have found that glossy photo paper for inkjet printers makes a very good toner transfer film **[6]**. Epson glossy photo paper gave particularly good results. I have also had good results from Xerox brand glossy photo paper.

Whichever brand of inkjet photo paper you decide to use, I would advise you to do a hot iron test to make sure it can survive the heat of the laser printer or copier.

TONER TRANSFER. If you are using Press-N-Peel or similar transfer film, you should ignore some of the advice below and follow the manufacturer's instructions. I printed the track pattern onto a sheet of Epson glossy photo paper. My artwork can be downloaded from

http://homepage.eircom.net/~ei9gq/art.html

Note that this track pattern has already been reversed so that is can be printed without modification. If you don't have a laser printer, the pattern can be printed on ordinary paper using an inkjet printer and photocopied on to the transfer film. The PCB should be thoroughly cleaned with fine steel wool and dried with a lint-free cloth. The photo paper or transfer film is then placed toner-side down on the surface of the copper PCB. One edge of the film can be secured with some paper masking tape.

The toner is then ironed onto the surface of the board using a hot clothes iron. The amount of time and pressure required will depend on the temperature of the iron and the type of toner used in the printer or copier. I use a hot iron at close to maximum setting. The iron is placed on the back of the photo paper and pressed firmly down for about 10 seconds while the toner sticks to the PCB surface. The iron is then moved slowly back and forth for a further 15 to 20 seconds. The amount of pressure required will depend to some extent on the complexity and size of the PCB pattern. Heavy pressure on the iron will tend to flatten the toner which will make lines slightly thicker and pads slightly larger.

Do not attempt to peel the photo paper off the PCB at this stage. The board should be placed in warm soapy water which will soften the photo paper. You should then unplug the iron and leave the room immediately. All homebrew radio enthusiasts have a natural curiosity and a tendency to poke at things, even when we know we shouldn't. Leave the board in the warm water for at least 20 minutes before removing the photo paper. By this time, the paper will be very wet and should be easy to remove. Any remaining paper residue can be brushed off with your fingertip while holding the board under running water.

Very stubborn residue can be removed with a coarse sponge. Don't be too afraid of damaging the track pattern, the toner is usually quite durable. Any small gaps in the toner can be repaired with an etchresist pen. **Photo 1.5** was the board after ironing.

ETCHING. The board was etched in a ferric chloride solution. Ferric chloride can be purchased in liquid or crystal form from most electronics suppliers. See the supplier's data-sheet for information about safe handling and proper disposal of ferric chloride. Ferric chloride is corrosive and quite toxic in concentrated form. However, it is probably no more dangerous than some household chemicals like bleach or drain cleaner. Wear rubber gloves and eye protection. Wear old clothes, ferric chloride stains won't wash out. Sodium persulphate is a good alternative etchant.

Etching time depends on the temperature, strength and condition of the etchant, the thickness of the copper foil and the area of copper to be removed. Commercial etching tanks use heated etchant and a stream of air bubbles or mechanical agitation to speed up the etching process. Boards that are upside down (with the copper facing downwards) will etch faster than boards that are lying copper side up. Double sided boards can be etched standing on edge if the etchant is deep enough.

Typical etching times range from five minutes for a heated/agitated tank up to about an hour for an unheated etching tank on a cold day. **Photo 1.6** shows the PCB in the etchant. Note that the PCB is floating face-down on top of the etchant. Also note the length of insulated wire that allows for

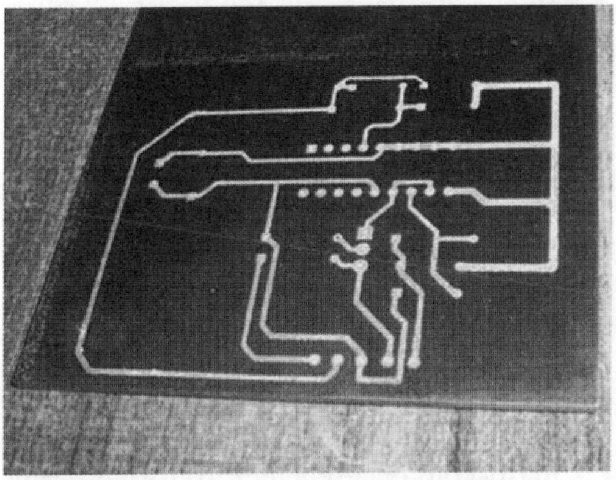

PHOTO 1.5: A PCB board after the artwork was imprinted on it using standard photo paper and an iron.

PHOTO 1.6: PCB in the etchant. Note that the etchant is both corrosive and toxic.

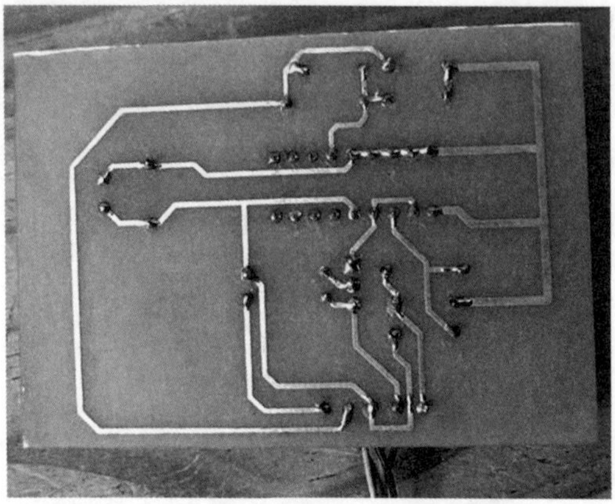

PHOTO 1.7. The bottom of our PCB showing the copper track and solder.

PHOTO 1.8: The top of our PCB showing the components.

PHOTO 1.9: A close up of the PCB showing the high quality that can be achieved using cheap and simple techniques.

handling of the board without coming into contact with the etchant. The board should be wiggled from side to side several times during the etching process. This board took about 35 minutes to etch. I would recommend that you use a deep plastic tub rather than the shallow trays that are supplied with etching kits.

Once the board is etched it should be rinsed with clean water. If you find that there are still a few traces of unetched copper, you can put the board back in the etchant for a few minutes before you rinse it again. After etching, the toner can be removed from the board with paint/varnish remover or steel wool.

DRILLING. Unlike hand-drawn PCBs, CAD-produced PCBs tend to have very small pads for component leads. I used a 0.75mm drill to make the holes in this PCB. 0.8mm is about the largest drill size that can be used with this PCB layout. I normally use HSS (high speed steel) drills for this kind of work. Tungsten carbide drills take much longer to wear out but they break easily.

ASSEMBLY. The board should be cleaned again with fine steel wool. If you want to make a posh looking PCB, you can print the top silk-screen layer onto your photo paper or transfer film and iron it onto the top of the PCB. Next the board should be stuffed with components and soldered. I use a small brass-bristled spark-plug cleaning brush to remove any residual flux. As well as removing the flux, this takes the shine off the solder joints. If I want to make a really posh job of it, I re-flow all of the solder joints with the tip of the soldering iron to restore the shine.

As this is a well tried and tested circuit and it is unlikely to need repair or modification, I gave the copper side of the board a couple of coats of clear varnish. This should keep it in mint condition for many years. The varnish is plasti-kote clear enamel from B&Q. This method of PCB production produces boards that are not quite as good as the photographically produced boards from commercial PCB makers, but as you can see from **Photo 1.9**, they come pretty close. Note that the board in Photo 1.9 is not the result of a long trial and error process. You can expect to make boards as good as or better than this on your kitchen table. The PCB is for a simple PIC programmer.

PIC PROGRAMMER COMPONENT LIST			
C1	100nF	D2	13V Zener
C2	100µF	D3	5.1V Zener
R1, R2, R4	10kΩ	Q1	BC547
R3	1k5	18 pin IC socket (preferably ZIF)	
D1, D4, D5	1N4148		

Producing a PCB using photographic methods is a homebrew version of the system employed by most commercial PCB manufacturers. The circuit is a simple microphone amplifier for an SSB transmitter. The schematic is shown in **Figure 1.3**.

ARTWORK. As we will be using a pre-coated PCB with a positive photoresist, we will need a positive master artwork (copper areas opaque). Note that some PCB systems use a negative photo resist and require a negative master artwork (copper areas transparent). The artwork was produced

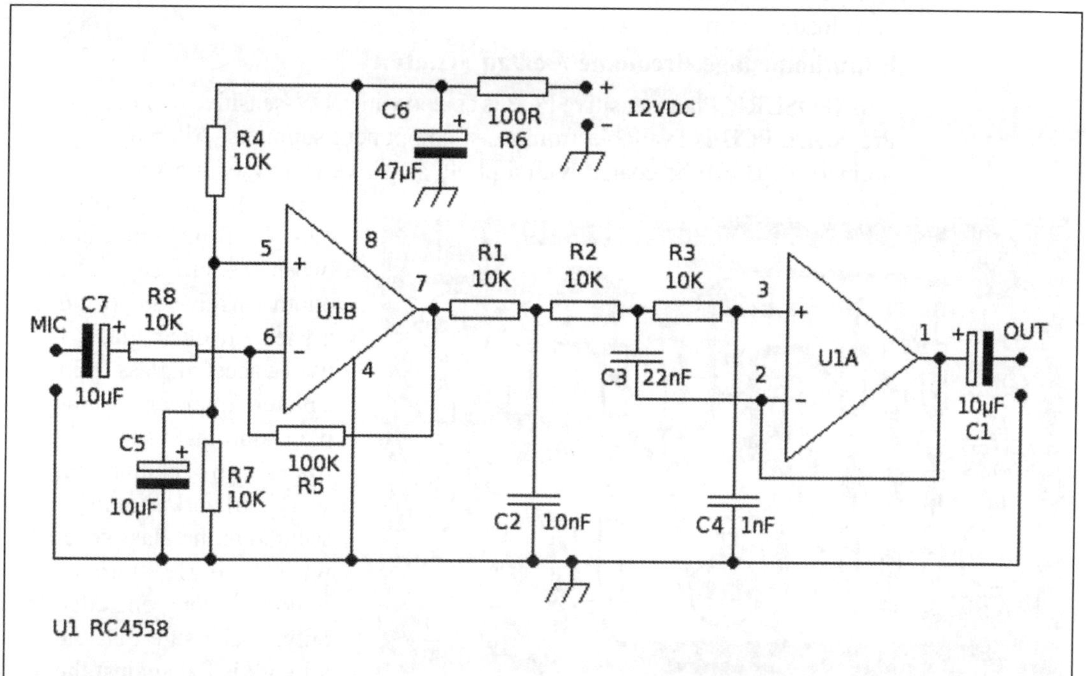

FIGURE 1.3: A schematic of a simple microphone amplifier for an SSB transmitter.

using PCB [7] CAD software. **Figure 1.4** shows the layout of the board. The copper track layer was printed on overhead projection film using a laser printer. If a laser printer is not available, the track pattern can be printed on white paper using an inkjet printer and photocopied onto the OHP film. Make sure you use the type of OHP film that is suitable for use in photocopiers and laser printers. The image should be printed on the side of the OHP film that is closest to the PCB. If the image is on the opposite side of the film, some UV light will creep around the edges of the opaque areas of the artwork. This will lead to thinner tracks and fuzzy edges. Do a test print on ordinary A4 copy paper before you print on the film. Hold the test print up to the light and look at it from the other side to check that the image is correct. You may find that you need to print a mirror image of the track pattern. Many High Street stationers will be able to photocopy onto OHP film. Remember to take an eight-pin IC socket with you so that you can check the size of the printed image.

High quality master artwork is the key to success. If the black areas are absolutely opaque, your PCB tracks will have sharp, clear edges and will be free of cracks and holes. You should set the printer/copier for maximum toner density. Contrast can be improved by stacking two sheets of film. This requires very careful alignment of the two sheets. The master artwork for this project was made from two sheets of laser OHP film. The track layout was printed using a 1,200 dpi HP laser printer with the toner density set to maximum. After printing, I cut off most of the surplus film around the printed area. One sheet was cut larger than the other so that I could use a strip of Sellotape to keep the sheets aligned. The alignment marks produced by most PCB software will help you to achieve perfect alignment. My artwork can be downloaded from

http://homepage.eircom.net/~ei9gq/art.html

EXPOSURE. Photosensitive PCB is coated with a UV-sensitive photoresist. Pre-coated PCB is available from most component suppliers. Alternatively, ordinary PCB can be coated with a photo sensitive layer (see hints and tips section). The artwork must be firmly held between the surface of the photo sensitive board and a sheet of clear glass. I used a sheet of glass from a picture frame and a simple clamp that was made from a strip of aluminium.

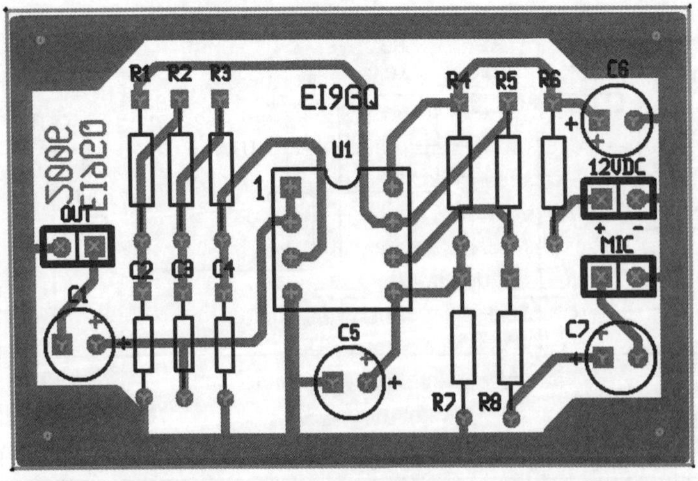

FIGURE 1.4: PCB layout for the microphone amplifier.

The artwork should be secured to the glass sheet with a single strip of Sellotape along one edge only. Make sure that the artwork is flat against the glass. If you are using double sheets, check that both sheets are properly aligned. The photo sensitive PCB should only be handled in subdued light. Once the protective film has been removed from the board, it should not be exposed to sunlight or bright fluorescent light. The board should be placed in contact with the toner side of the artwork and clamped firmly in place. **Photo 1.10** shows my home made exposure frame.

PHOTO 1.10: The homemade exposure frame.

After checking the alignment of the PCB and artwork one last time, the PCB should be exposed to a UV light source. I have a home-made UV light box that I use for exposing PCBs. Since most *RadCom* readers will not have a UV light box, I decided to try some of the alternatives.

The board can be exposed by placing the exposure frame in direct sunlight. The optimum exposure time will depend on the brightness of the sun, the thickness of the glass and the type of photoresist on the PCB. My first attempt was a complete failure. The board was overexposed and the tracks almost completely disappeared when the board was developed. The second attempt was much better. I exposed the board to direct sunlight for five minutes. The board was still very slightly overexposed but the end result was an almost perfect PCB. The best results were obtained with an exposure time of just two minutes.

After reading a few reports from successful PCB makers who use quartz-halogen lamps as a UV light source, I bought a 500W halogen light and mounted it on a shelf. The PCB exposure frame was placed a safe distance away from the front of the lamp. At a distance of 28cm from the glass cover of the lamp, the exposure frame gets quite warm but not too hot. For the first attempt, I used an exposure time of eight minutes. To my great surprise and delight, this produced a perfect PCB. The board needed almost five minutes in the developer before the unwanted etch resist was washed away. This is a little longer than normal, which would indicate that the board was slightly underexposed. For the next attempt, I increased the exposure time to ten minutes. Once again, the PCB was absolutely flawless and the developing time was a more reasonable three to four minutes. I have exposed several boards using the halogen lamp and so far I have enjoyed a 100% success rate.

Ultra violet fluorescent tubes of the kind used for disco lighting (black light), banknote scanners, suntan lamps or insect zappers can be used for PCB exposure. Ordinary fluorescent tubes can also be used, but exposure times will be much longer. A single Philips 30W TLD 30W/33 tube gave good results with an exposure time of one hour at a distance of 10cm. DON'T use EPROM eraser or germicidal UV tubes, because they produce short wavelength UV that is harmful to eyes and skin.

DEVELOPING. The two most commonly used PCB developers are sodium hydroxide and sodium metasilicate. Sodium hydroxide is widely available from most hardware stores as caustic soda or drain cleaner. Both types of developer can give excellent results. I have found that caustic soda is a bit more difficult to use. To get good results, the strength and temperature of the caustic solution must be carefully controlled. Sodium metasilicate-based developer is much less critical and will deliver consistent results over a wider temperature range. The developer used for the microphone amplifier boards was made by dissolving one teaspoon (5gm) of sodium hydroxide crystals (B&Q own brand caustic soda) in 500ml of water. You must resist the urge to add a little bit extra for good measure. This solution makes a good PCB developer at a typical room temperature of 17-20°C. Sodium metasilicate developer is available in liquid or crystal form from most electronic component suppliers. You should follow the supplier's instructions for mixing and safe handling.

The boards were developed in a plastic margarine tub (see **Photo 1.11**). The developer should be rocked very gently from side to side while the etch resist is washed away. You should wear eye protection while you are developing the board. You really don't want to get caustic chemicals in your eyes. After about one minute in the developer, the outline of the PCB tracks appears on the board. After two minutes, you should start to see bare copper. At three minutes, the board should be almost completely developed. The board should be left in the developer for a further minute to ensure that all of the exposed etch resist is washed away. The board should now be removed from the developer and thoroughly rinsed with running water. If your developing time is longer than 4-5 minutes, you probably need to increase the UV exposure time or slightly increase the strength of the developing solution. The remaining photo resist should be an exact replica of the master artwork. The board

PHOTO 1.11: The margarine tub used for developing boards.

PHOTO 1.12: Etching tub. Note the wire aro-und the board for safe handling of the board.

should be handled very carefully at this stage. The resist must not be scratched before the board is etched.

ETCHING. See previous section for details of how to etch a PCB using ferric chloride as an etchant. I have always used ferric chloride as a PCB etchant in the past, so I have no experience of using any other etchant. Just for the sake of trying something different, I used a sodium persulphate etchant for this project. Sodium persulphate is available from Maplin and many other suppliers. I mixed 100g of sodium persulphate with 500ml of warm water as specified in the supplier's instructions. When the powder was fully dissolved, I placed the PCB in the warm etchant. **Photo 1.12** shows the etching process. Note the insulated wire which allows for handling of the board without coming into contact with the etchant. The board was fully etched in 13 minutes. Sodium persulphate works best when it is warm. A heat lamp can be used to warm the etchant from above. I used a fridge thermometer to monitor the temperature of the etchant and the halogen lamp to provide the heat. Be very careful about how you do this. A melted container of etchant would make a really horrible mess. A glass container would be safer than my plastic tub. You need to stay close enough to monitor the temperature, but you should avoid breathing the vapour from the warm etchant. The board should be wiggled gently from side to side several times during the etching process. This helps to clear etched copper from the surface of the board.

After etching, the board should be rinsed thoroughly under running water. A quick splash under the tap is definitely not good enough. When you are ready to drill the board and populate it with components, the etch resist can be stripped off the board with paint/varnish remover or very fine steel wool. Alternatively, the board can be re-exposed to your UV light source without using the artwork as a mask or left in bright sunlight for a few minutes. The etch resist can then be removed by putting the board back in the developer for a few minutes. If you are not planning to use the board for a while, the etch resist can be left on the board as a protective coating.

DRILLING. Hand-cranked egg-beater style drills are just about usable for hand-drawn PCBs. To drill the tiny holes required by CAD produced PCB layouts, however, a 12V miniature PCB drill or a mains powered rotary tool will do a much better job.

After many years of breaking drill bits, I have at last discovered a reliable way of drilling holes in PCBs. I use a 12V PCB drill with a reduced supply voltage of around 6 volts. Using a lower supply voltage reduces the kick caused by the torque reaction when the drill starts turning. The lower voltage also reduces the speed of the drill, which results in longer life from HSS drill bits. I use a push-to-make foot switch to operate the drill. This leaves both hands free to hold the drill and PCB. You should wear safety goggles. If you don't have any, go and get some now. To avoid drilling holes in the table, you should drill the board on an old telephone book or a paperback.

The exact procedure I use is as follows: Place the tip of the drill in the centre of the pad to be drilled. I use a bright light so that I can see what I'm doing. Taking care to keep the drill absolutely vertical, press the foot switch.

PHOTO 1.13: The PCB board after drilling.

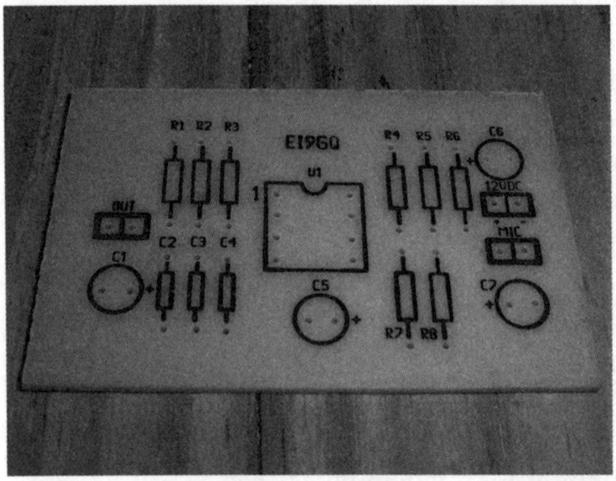

PHOTO 1.14: The silk-screen layer from the cad software printed on the board.

Don't hold the drill too firmly. The hole in the centre of the pad will act as a guide for the drill. Even if you are a little off centre, the drill will tend to centre itself with the hole in the copper pad. When the hole has been drilled, withdraw the drill vertically while it is still turning and release the foot switch. I used a 0.75mm drill for this PCB. **Photo 1.13** shows the drilled board.

Most constructors recommend using a drill stand for PCB work. I prefer to use a hand-held drill and the above technique. I have drilled several hundred holes during the last couple of weeks without breaking a single drill bit or drilling an off-centre hole.

Because most of my PCBs end up in a screened box with the lid screwed on, I don't really care how they look. This time I printed the top silk-screen layer from the CAD software on the top of the board (see **Photo 1.14**). The pattern was printed on glossy photo paper and ironed onto the top of the board. The same warnings apply to this process as I detailed previously.

DOUBLE-SIDED BOARDS. There are a few different ways of making double sided boards using photographic methods. It is possible to process both sides of the board at the same time by using a double sided exposure frame and a pair of UV light sources. This requires extremely accurate alignment of the artwork on both sides. I find it is a lot easier to do one side at a time. The first side of the board is exposed, developed, etched and drilled as it would be for a single sided board. Use a sharp new drill to reduce the risk of damaging the photoresist and protective coating on the opposite side. When the first side is finished, it can be covered with strips of PVC tape to protect it while you are etching the second side. Now you can remove the plastic protective film from the second side of the board. The artwork for the second side should be aligned carefully with the holes in the board. This ensures that both sides of the board are properly aligned. The second side of the board can now be exposed, developed and etched. Be very careful to protect the first side from the etchant while you are etching the second side. After etching, the protective tape and photoresist can be removed, hopefully to reveal a perfect double-sided PCB.

If you are really paranoid about protecting the first side of the board, you can give it a coat of paint or varnish before you start work on the second side. The paint can be removed with paint/varnish remover after the second side is etched. I have found that using a few strips of PVC insulating tape is just as effective.

SURFACE MOUNT. SMT (surface mount technology) boards as shown in **Photo 1.15** are made using exactly the same methods as boards for leaded components. PCBs with leaded components are often referred to as PTH (plated through hole) technology. This is a bit of a misnomer in our case because most amateurs will not have the equipment to make boards with electroplated holes. I have used the photographic methods described here to make boards using surface-mount ICs and 603-type chip capacitors and resistors. The main differences between boards for leaded components and boards for SMT are the methods used for mounting and soldering the components.

PHOTO 1.15: here is the same microphone amplifier circuit built using surface mount technology (SMT).

HINTS AND TIPS. As an alternative to pre-coated PCB, you can apply a coat of photo resist to ordinary PCB laminate. The photo resist liquid can be sprayed or wiped onto the surface of the PCB. Extreme cleanliness is required for this process. See Harry's Homebrew Homepage **[8]** for a good tutorial.

Double-sided boards can also be made by etching a pair of single-sided boards and sticking them together with epoxy resin. If you use 0.8mm laminate, the finished product won't be much thicker than standard 1.6mm board.

For small circuit boards, you can get more mileage per sheet of film by printing two copies of your layout on each sheet. I find this useful because I usually use a double layer of film for the artwork. This way I get two for the price of one.

Large unused areas on your PCB should be filled with grounded copper. Apart from the electrical advantage of having a better ground plane, you will use a lot less etchant and developer.

Not all PCBs are created by chemical etching. Some PCB prototyping systems use a cutting tool to remove unwanted copper from the board. At least one of the regulars on the 80m homebrew net (3,727kHz) makes all of his PCBs with a home-made PCB cutting/milling machine.

Once again, I will remind you to wear eye protection. Remember the old saying "It's all fun and games until someone loses an eye".

READERS' HINTS AND TIPS. Many thanks to the readers who have sent me their comments and suggestions. The PCB-making techniques previously covered in *RadCom* prompted some readers to share their experiences.

James, G3RUH, describes his method of making PCBs, "Another effective method is to use 'PCB Transfers'. See RS part number RS 436-083. These are press-on shapes reminiscent of Letraset.

1. Print your artwork at 1:1 onto paper and check with a ruler that it's correctly sized!
2. Sellotape the artwork to the copper-clad board.
3. Use a fine punch to dot all the holes through the paper artwork.

4. Now use the PCB transfers to draw the artwork onto the PCB. Do the ICs first, then the component pads, and then the tracks. Regularly burnish the tracks etc using a finger nail, with a piece of the semi-transparent protective interleave sheets as a buffer. Touch up with etch-resistant pen.
5. Etch.
6. Remove artwork with wire wool etc."

Ted, G3NPL, reports that he gets excellent results when he uses his inkjet printer to print PCB artwork onto overhead projection film. I haven't enjoyed much success with my own old Canon i350 inkjet. Obviously, the newer photographic printers are better suited to the job. Ted writes: "I have been using an inkjet of the newer type for printing photographs, in this case a CANON PIXMA iP2000. It gives excellent results on all boards. It must be set to high resolution and with overhead film set to high resolution paper."

Alan, G4GQL, sends this gem of a tip for improving the contrast of laser printed or photocopied artwork. Alan writes: "Lay the positive on a flat table over a piece of paper bigger than the positive. With a piece of cotton wool dipped in white spirit carefully wipe over the image - it will turn jet black. Try it and hold it up to the light - you will be amazed how dense it is. You must use a good white spirit. And dry the film in front of a fan heater."

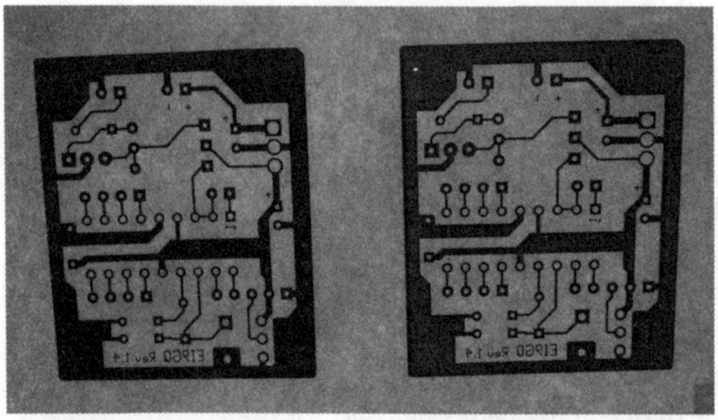

PHOTO 1.16: Homebrew reader Alan, G4GQL uses a clever technique to improve the contrast of laser-printed or photocopied artwork.

I tried this technique and, as Alan promised, I was amazed at the result. **Photo 1.16** shows an OHP transparency with two identical PCB track patterns printed on it. I won't tell you which side has been treated with white spirit. Alan recommends a fan heater for drying the film. I used a 1,400W hair-dryer. Please be careful when you are drying the film. A powerful heater and a bottle of white spirit could be a very dangerous combination.

EQUIPMENT USED

Small photograph frame (for the glass)
500W halogen lamp (B&Q)
HP LaserJet 1200 printer
Pre-coated photo sensitive PCB Maplin JP56
Staedtler Lumocolor transparent photocopy/laser film (NOT inkjet type)
Sodium persulphate etchant 100g Maplin NO6CG
Sodium hydroxide. (B&Q caustic soda)
Photographic or fridge thermometer
12V PCB mini-drill
Assorted plastic tubs and a roll of Sellotape

MICROPHONE AMPLIFIER COMPONENT LIST			
R1, R2, R3, R4, R7, R8	10kW	C3	22nF
R5	100kW	C4	1nF
R6	100W	C6	47µF electrolytic
C1, C5, C7	10µF electrolytic	U1	RC4558 or similar dual op-amp (TL072, LF353)
C2	10nF		

Metal Bashing

A lot of my radio projects are housed in commercially made metal boxes. Many component suppliers have a good range of enclosures available at a reasonable price. Aluminium sheet has become so expensive that it is often cheaper to buy a simple aluminium box than it is to make one. The more rigid die-cast alloy boxes are quite expensive, but they are ideal for projects where mechanical stability and good screening are important. For larger projects, enclosures that have a vinyl coated steel lid and aluminium bottom half are easy to work with and very good value. It is often possible to re-use enclosures from scrap equipment to house your projects. The steel enclosure used for the power supply project was recovered from a consignment of scrap medical equipment.

Heavy steel enclosures are quite difficult to work with. Cutting out rectangular or round holes in sheet steel can be quite a challenging task when you only have a few hand tools. I still have nightmares about the beautiful shiny stainless steel boxes I acquired from a local scrap yard. I couldn't cut them with a hacksaw or even drill holes in them!

When you find that none of the available enclosures is suitable for your project, it is often easier to make a new box from scratch than it is modify an existing one. I built the enclosure for my HF transceiver **[9]** because I couldn't find a commercially made one that was suitable for the job and there was nothing suitable in the scrap yard. I have also built special enclosures for VHF/UHF helical and cavity filters. The stripline enclosure of my VHF linear amp is made from a sheet of 1mm aluminium.

I have very little experience of working with sheet metal and very few tools. My current metalworking kit consists of a blunt hacksaw, a power drill with a few HSS bits (mostly blunt or broken), an electric jigsaw, a few simple hand tools, hammer, screwdrivers, pliers etc. A strong pair of scissors can be used to cut aluminium sheet up to a thickness of about 1mm. The two most useful tools in my metalworking kit are a sharp carpet knife and a small nibbling tool. The least useful tool is the jigsaw which is far too fast and powerful for delicate work.

SAFETY. Always wear safety goggles when working with power tools. Using tools for purposes other than the purpose they were designed for can be particularly dangerous.

Apart from the fact that you will be using a much stronger pair of scissors or shears, the process of making an enclosure is much the same as making a new suit of clothes. You will need to decide what size box will be required, make a pattern for the various panels, cut the panels from sheet metal and form them into the required shape. The design of the enclosure will depend on your own individual requirements. The enclosure can be just a simple box or a more elaborate structure like a chassis with circuits mounted on both sides. Many homebrew projects call for an enclosure which is divided into separate compartments to provide improved screening of the enclosed circuit.

When you have decided on a layout and size for your project, you should write down the exact measurements of your enclosure and spend some time drawing up a detailed pattern for the metal panels. The pattern can be drawn by hand on graph paper or with a computer CAD (computer aided design) or drawing program. I use a simple vector drawing program called Xfig on a Linux PC. Similar software is available for Windows and Mac OS. The pattern should be drawn to scale. Once you have a detailed pattern of the panels, you can cut them from sheet metal. I usually use 1mm aluminium sheet for my homebrew projects because it is relatively easy to cut and bend. I will assume that like me, you don't have access to proper metalworking tools like a guillotine or bending brake.

I have tried using an electric jigsaw for cutting sheet aluminium, but I found that I could never manage to cut a straight line. The jigsaw also tended to leave a slightly jagged edge on the cut. A simple 12in hacksaw is a bit slower, but it usually does a better job. There is a much better way of cutting straight edges in aluminium sheet. If you use a very sharp knife to cut a deep groove in the surface of the aluminium sheet, you will find that it is quite easy to bend the sheet along the line of the cut. The sheet can be bent around a sharp metal edge like a length of angle iron or the edge of a metal bench. Smaller pieces can be bent in a bench vice. Once the sheet has been folded back through 180 degrees, it is very easy to break off the bent section. This method gives a perfectly straight cut every time. A steel rule will help you to draw a straight line and a Stanley or similar type of carpet knife can be used to cut into the aluminium sheet. Be careful not to ap-

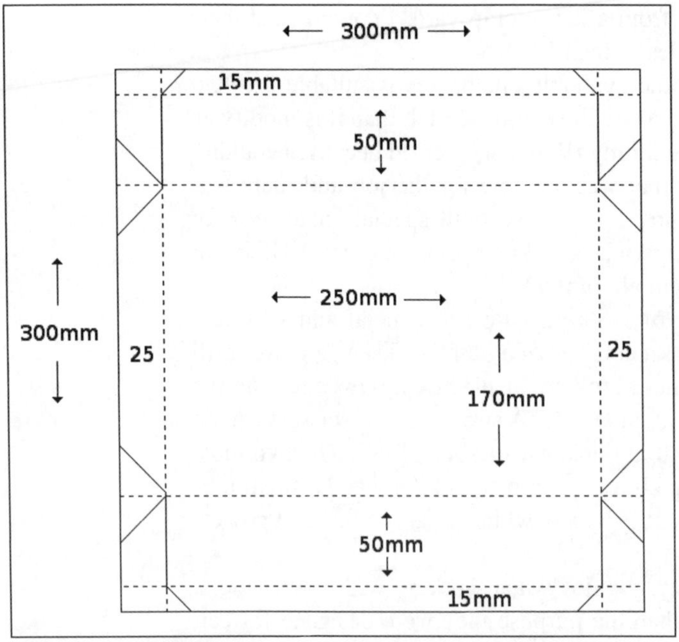

FIGURE 1.5: Template for the base of the box.

ply too much pressure to the carpet knife in case it breaks. A metal bodied carpet knife is much safer than plastic types. Retractable knives with plastic bodies and renewable blades with snap-off sections are not suitable for this job.

You can use a similar technique to form nice clean 90° bends in your aluminium sheet. If you use the carpet knife to score along the line of the bend, it will help you make a nice straight and sharp bend. The line should be scored on the outside radius of the bend. You should only use the knife on one side of the sheet. Don't cut too deep or the sheet might tend to break along the cut.

References

[2] www.k7qo.net

[3] *VHF/UHF Manual* Chapter 3. G.R. Jessop, G6JP. RSGB 1983

[4] http://pcb.sourceforge.net

[5] PIC-A-STAR Part 3. *RadCom* October 2002.

[6] http://homepage.eircom.net/~ei9gq/pcb.html

[7] http://pcb.sourceforge.net

[8] http://web.telia.com/~u85920178/data/pcb2.htm

[9] http://tinyurl.com/ygnfdg

[illegible]